International Conference on Advanced Materials, Microscopy and Energy (ICAMME 2019)

IOP Conference Series: Materials Science and Engineering
Volume 783

Meknes, Morocco
15 - 19 April 2019

ISBN: 978-1-7138-2667-5
ISSN: 1757-8981

Printed from e-media with permission by:

Curran Associates, Inc.
57 Morehouse Lane
Red Hook, NY 12571

Some format issues inherent in the e-media version may also appear in this print version.

Institute of Physics
Dirac House, Temple Back
Bristol BS1 6BE UK

Phone: 44 1 17 929 7481
Fax: 44 1 17 920 0979

techtracking@iop.org

Additional copies of this publication are available from:

Curran Associates, Inc.
57 Morehouse Lane
Red Hook, NY 12571 USA
Phone: 845-758-0400
Fax: 845-758-2633
Email: curran@proceedings.com
Web: www.proceedings.com

TABLE OF CONTENTS

MATERIALS ENGINEERING AND ENVIRONMENT

ADVANCED MATERIALS AND ENERGY

IOP Conf. Series: Materials Science and Engineering **783** (2020) 011001 doi:10.1088/1757-899X/783/1/011001

International Conference on
Advanced Materials, Microscopy and Energy (ICAMME'19)
April 15-19 / 2019, Meknes, Morocco

Preface

The International Conference on Advanced Materials, Microscopy and Energy (ICAMME'19) was held in Faculty of Sciences University Moulay Ismail Meknes-Morocco from April 17 to 19, 2019. This event has included different kinds of presentations given by researchers and experts from the national and international scientific community, including keynote speakers, special sessions, posters and tutorials. It has covered a wide spectrum of topics.

A specialized spring school was held from April 15 to 16, 2019 before the Conference. The lectures during this school have covered fundamental and applied aspects related to Advanced Materials and Advanced Microscopy and spectroscopy techniques. The lecturers were international experts in these areas. The school was intended in order of priority to PhD students, Masters 2, Engineers and researchers.

ICAMME'19 was an ideal platform for all International and National Scientists, Professors, Students, and Industrials. There were feature talks by eminent personalities from academics and industries on recent advances in field of Materials Science, Microscopy, Engineering, Technology and Energy. It has known more than 150 participants around the globe with thought Keynote lectures, Oral Presentations and Poster Presentations. This was an excellent opportunity for the delegates from Universities and Institutes to interact with the world class Scientists.

The conference ICAMME'19, over three days, has permitted:

1. *to round up specialists in the advanced materials research field (theoretical and experimental),*
2. *to discover the scientific community and research in the Moulay Ismail University,*
3. *To review to the younger generation today, through various scientific presentations, the evolution of the advanced materials, and the revolution in nanotechnology and their applications in renewable energy.*

We thank the members of the Scientific Committee and all the invited speakers.

We recall that the financial support of the conference was provided largely by:

1. *The Moulay Ismail University*
2. *The Laboratory of Advanced Materials Studies and Applications (LEM2A)*

We are sure that everyone has found in this meeting an important topical interest, a great pleasure on exchanging with the inter-Mediterranean scientific community.

Prof. Abdelhai Rahmani,
General Chairman

The scientific content of this volume of abstracts falls under the responsibility of the authors.
The organizing committee thanks the scientific committee for the critical reading of the abstracts and the selection of papers.
ICAMME2019

IOP Conf. Series: Materials Science and Engineering **783** (2020) 011002 doi:10.1088/1757-899X/783/1/011002

Peer review statement

All papers published in this volume of *IOP Conference Series: Materials Science and Engineering* have been peer reviewed through processes administered by the proceedings Editors. Reviews were conducted by expert referees to the professional and scientific standards expected of a proceedings journal published by IOP Publishing.

IOP Conf. Series: Materials Science and Engineering **783** (2020) 012001 doi:10.1088/1757-899X/783/1/012001

Printing temperature effects on the structural and mechanical performances of 3D printed Poly-(phenylene sulfide) material

A El Magri*[1, 2], S Vaudreuil[1], K El Mabrouk[1] and M Ebn Touhami[2]

[1] Euromed research center, Euromed Engineering Faculty, Euro-Mediterranean University of Fes, B.P. 51, Meknes Road, Fes, Morocco.
[2] Laboratory of Materials Engineering and Environment: Modeling and Application, Faculty of Sciences, Ibn Tofail University, PO Box 133, 14000 Kenitra, Morocco.

*Email:a.elmagri@ueuromed.org

Abstract. Fused Deposition Modeling (FDM) is the most popular and widely used additive manufacturing techniques for plastic materials. In FDM, the Z-height of the object is achieved by depositing extruded polymer layer upon layer. To reduce the time and cost of printed parts requiring specific properties, many processing parameters can be adjusted to optimize the FDM process. Among those is the nozzle temperature. In this study, Poly-(phenylene sulfide) (PPS) was chosen because of its mechanical performance, making it attractive for high-performance applications. In this work, the impact of printing temperature on the mechanical and structural properties of printed parts was carefully tested. Results show that a 340°C printing temperature yield the parts with the highest tensile properties, with a degree of crystallinity superior to as-received PPS.

1. Introduction

Additive Manufacturing (AM) has shown the possibility of significantly reduce both time and cost of product development. AM, per the ASTM F2792-12a definition, is defined as "a process of joining materials to make objects from 3D model data, usually layer upon layer, as opposed to subtractive manufacturing methodologies". AM is also called, on a more familiar basis, 3D printing even though it doesn't reflect the full potential of what AM can achieve. This new integrated technology enable the production of complex geometries hard to impossible to achieve with conventional methods [1]. Thanks to its ease of use and lower investment and operating costs, Fused Deposition Modeling (FDM) [2][3] has become the most used 3D printing process. FDM operates on the basis that a thermoplastic filament is extruded, layer by layer, in its viscous state through a heated nozzle.

Literature abound with published works on AM of commodity polymers [4] such as polycarbonate (PC), polylactic acid (PLA), acrylonitrile butadiene styrene (ABS) and PC–ABS blends [5]. For high-performance thermoplastics such as PPS in AM, literature is almost nonexistent. While FDM is more user-friendly in its operation, results can be significantly affected by many parameters, including printing temperature and speed, layer thickness and raster orientation. Multiple works have concluded that mechanical properties will depend on those parameters [1], [6]–[10]. Kishore et all [11] have used two high performance semi-crystalline polymers (PEEK and PPS) with FDM process to evaluate the relationship between rheological properties and processing conditions. The authors have shown that a rheological analysis is helpful in identifying appropriate FDM processing conditions and key control parameters.

The present study aims to fill some gaps in published literature by investigating the relationship between printing temperature, mechanical and structural properties of PPS parts prepared by FDM process.

2. Materials and methods

The Poly (phenylene sulfide) (PPS) filaments used were purchased from FILOALFA (Italy) and processed using an INTAMSYS (Intelligent Additive Manufacturing Systems) FunMAT HT. This FDM system, having a build volume of 260 x 260 x 260 mm, is equipped with a heated bed and heated build chamber. As the melting temperature of PPS is 280°C, print head temperature above 300°C are required to achieve some flowability. All tensile bars used were printed directly on the heating bed, using a brim to increase adhesion and lower peel off risks. Infill was set to 100 % for all specimens to obtain solid-like samples. The major printing parameters used are listed in Table 1.

Table 1. FDM process parameters.

Parameters	Values	Units
Printing temperature	320-330-340-350	(°C)
Bed temperature	65	(°C)
Chamber temperature	30	(°C)
Infill line directions	[45/-45]	(°)
Layer height	0.15	(mm)
Line width	0.4	(mm)
Infill density	100	%
Number of bottom / top layers	6 / 6	layers
Number of perimeters	2	wall

Thermal and structural properties of PPS were studied by Differential Scanning Calorimetry analysis (DSC Q20 with RCS90 cooling system, TA Instruments). Samples were heated from 25 to 350 °C under a 50 mL/min nitrogen flow before being cooled back to -5°C. A heating / cooling rate of 10°C/min was used for all tests. Glass transition, melting temperatures and enthalpies were calculated from the endothermic and exothermic peaks, using the TA Universal analysis software. The degree of crystallinity (χc) was calculated from the first heating as follows:

$$\chi_c(\%) = \frac{\Delta H_m - \Delta H_{cc}}{\Delta H_m^{100\%}} \times 100 \qquad (1)$$

Where ΔH_m and $\Delta H_m^{100\%}$ are respectively the melting enthalpy of PPS and 100% crystalline PPS, ΔH_{cc} is the cold crystallization enthalpy. The melting enthalpy of a totally crystalline PPS material is 80 J/g [12].

Tensile properties were established using a Criterion C45.105 electromechanical universal testing machine (MTS, USA) equipped with a 10 kN load cell, see Figure 1(A). A crosshead displacement speed of 0.1 mm/s was used to apply load through self-tightening jaws. A data acquisition software (MTS TestSuite) was used to collect displacement (mm) and force (N) required for the tensile strain-stress curves and calculate the Young's Modulus and maximum tensile strength. For each printing temperature, six specimens were tested to increase accuracy. Figure 1(B) illustrate the tensile bar geometry used. This bar geometry was designed with dimensions divisible by 0.15 mm, which is the layer height used, to avoid creating missing or additional layers during slicing.

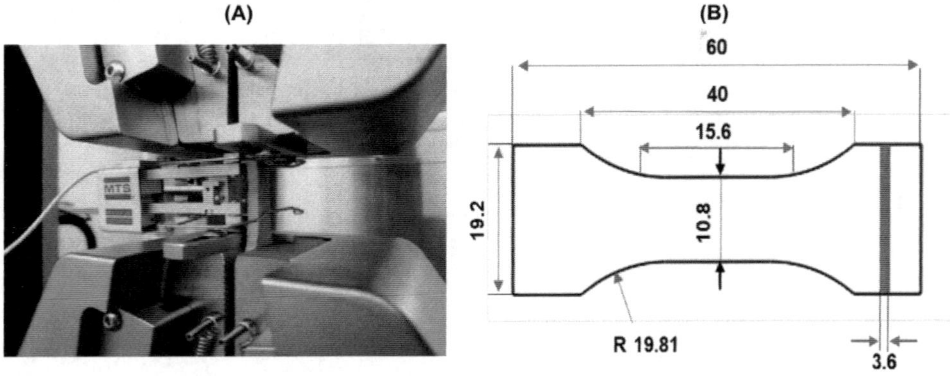

Figure 1. (A) MTS machine during tensile test,(B)geometrical illustration of tested specimen.

3. Results and discussion

3.1. Effect of printing temperature on thermal and structural properties

Thermal characteristics of PPS at various printing temperatures were established by differential scanning calorimetry (DSC) analysis. Figure 2 corresponds to the PPS Thermogrammes between 20°C to 330°C, using a 10°C/min heating/cooling rates. Four thermal transitions can be seen for PPS. During the heating phase, the first transition at ~90°C is the glass transition process where PPS change to a high elastic state. The exothermic transition at ~125°C is the cold crystallization, generally caused by an ordering of molecular chains in the crystalline PPS lamellae and associated to increased mobility throughout heating cycle [13]. The endothermic transition at 280°C amounts to the melting temperature. During cooling, only an exothermic transition at 224°C is observed which is associated to the crystallization of PPS. This last transition is caused by the rearrangement of molecular chains in the crystalline PPS lamellae due to increased mobility during heating. The first heating cycle in Figure 2 confirm that PPS is a semi-crystalline material.

It should be noted that printed PPS at 340°C exhibit a higher degree of crystallinity (37.32%) than the as-received polymer (Table 2). This higher crystallinity could be attributed to the creation of highly ordered lamella during printing at 340°C. Printing PPS at temperature approaching its degradation temperature will results in polymer chains break. The shorter chains will make more difficult the formation of crystalline lamella formation.

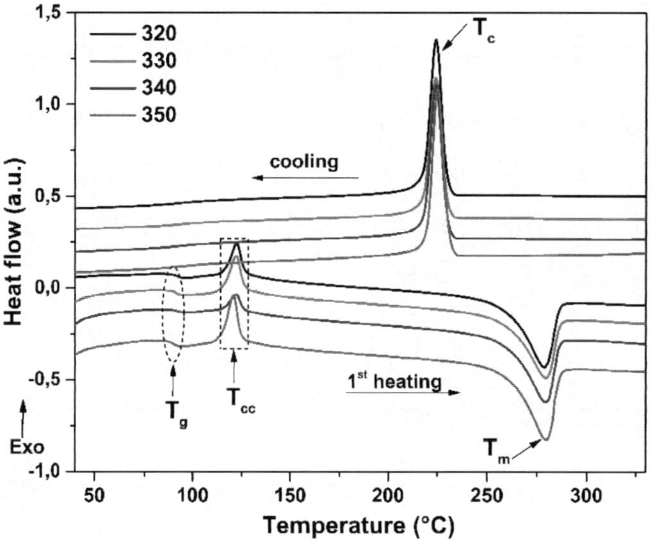

Figure 2. DSC curves of PPS after FDM printing at different printing temperatures.

Table 2. DSC analysis results.

Material	T (°C)	T_g (°C)	T_{cc} (°C)	T_m (°C)	ΔH_m (J/g)	ΔH_{cc} (J/g)	T_c (°C)	χ (%)
As received	-	92.3	122.9	279.6	33.3	14.8	223.9	23.0
After printing	320	91.2	122.3	279.2	31.1	9.49	224.0	27.0
	330	91.3	121.9	279.7	33.6	10.9	223.9	28.3
	340	91.2	122.3	280.0	35.6	5.78	224.0	37.3
	350	89.2	120.5	279.6	39.7	12.8	225.2	33.6

3.2. Effect of printing temperature on tensile properties

In order to evaluate the effects of printing temperature on the structural properties of printed parts, the behavior of the polymer system must be understood. Previous works reported that printing temperature affect significantly the flowing behavior and solidification process of printed filament [14]–[16] This will in turn influence the resulting properties during cooling phase such as rasters bonding and surface quality.

Figure 3 shows the variation of Young's Modulus and tensile strength with printing temperature from 320°C to 350°C. The enhancement of Young's Modulus and tensile strength can be observed with increasing printing temperature until a maximum is reached at 340°C, yielding values of 2030.58 MPa and 47.34 MPa respectively. This increase in mechanical properties with temperature can be explained by the stronger adhesion between rasters which will result automatically in an improvement of tensile properties.

These results illustrate the tendency of tensile behavior to follow the crystallinity degree. When printing temperature is raised from 340°C to 350°C, tensile properties begin to decline, with crystallinity decreasing from 37.3 to 33.6 %. This could be explained by the progressive collapse of the build layers and molecular chains deterioration.

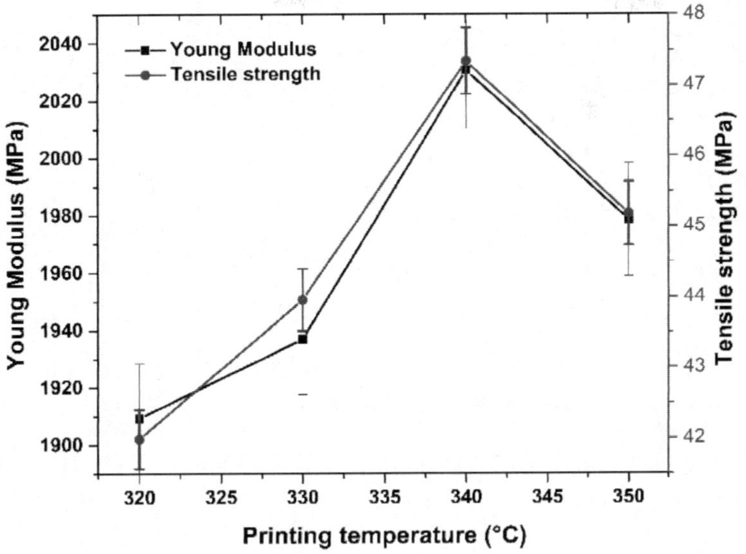

Figure 3. Printing temperature effect on tensile properties.

High printing temperature (350°C) tend to generate parts with more porosity and worst surface quality, things that negatively impact tensile properties (see figure 4).

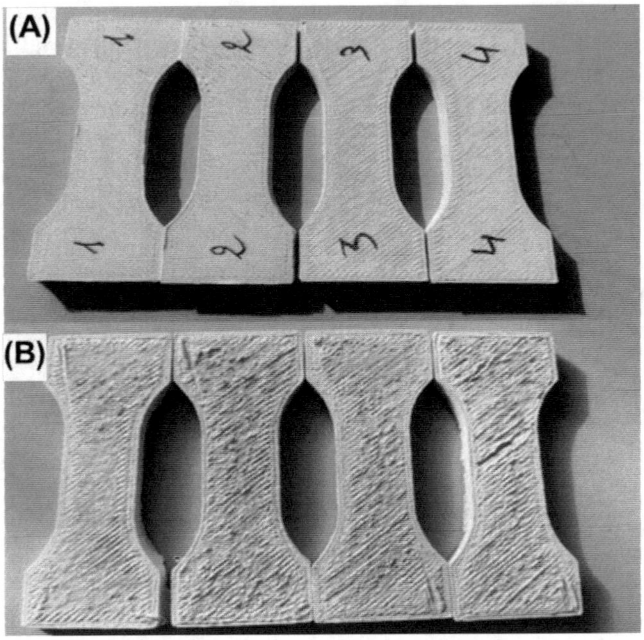

Figure 4. Effect of printing temperature on printed parts qualities (A) 340°C,(B) 350°C.

4. Conclusion

The objective of this work was to study the effect of the most critical FDM parameter, printing temperature, on final properties of printed parts. In order to investigate the relationship between printing temperatures and final properties of printed parts, temperatures between 320 and 350°C were tested. It was found that crystallinity and tensile properties were strongly affected by printing temperature, with a narrow window yielding the best values.

Acknowledgments

The authors also gratefully acknowledge the support from Euro-Mediterranean University of Fes, and the financial contribution of the Académie Hassan II des Sciences et Techniques and Safran Composites (France).

References

[1] Ahn S, Montero M, Odell D, Roundy S and Wright P K 2002 Anisotropic material properties of fused deposition modeling ABS *Rapid Prototyp. J.* **4** 248–257.

[2] Martínez J, Diéguez J L, Ares E , Pereira A , Hernández P and Pérez J A 2013 Comparative between FEM models for FDM parts and their approach to a real mechanical behaviour, *Procedia Eng.* **63** 878–884.

[3] Wu W , Geng P, Li G , Zhao D , Zhang H and Zhao J 2015 Influence of Layer Thickness and Raster Angle on the Mechanical Properties of 3D-Printed PEEK and a Comparative Mechanical Study between PEEK and ABS, *Materials* **8** 5834–5846 .

[4] Parandoush P and Lin D A 2017 review on additive manufacturing of polymer-fiber *composites, Compos. Struct.* **182** 36–53.

[5] Coey J M D 2012 Permanent magnets: Plugging the gap, *Scr. Mater.* **67** 524–529.

[6] Ziemian C , Sharma M and Ziemi S 2012 Anisotropic Mechanical Properties of ABS Parts Fabricated by Fused Deposition Modelling, *Mech. Eng. J* .

[7] Durgun I and Ertan R 2014 Experimental investigation of FDM process for improvement of mechanical properties and production cost, *Rapid Prototyp. J.* **20** 228–235.

[8] Sood A K, Ohdar R K and Mahapatra S S 2010 Parametric appraisal of mechanical property of fused deposition modelling processed parts, *Mater. Design.* **31** 287–295.

[9] Sood A K, Ohdar R K and Mahapatra S S 2012 Experimental investigation and empirical modelling of FDM process for compressive strength improvement, *J. Adv. Res.* **3** 81–90.

[10] Lee B H, Abdullah J and Khan Z A 2005 Optimization of rapid prototyping parameters for production of flexible ABS object, *J. Mater. Process. Technol.* **169** 54–61.

[11] Kishore V , Chen X , Ajinjeru C , Hassen A A , Lindahl J , Failla J , Kunc V and Duty C 2016 Additive Manufacturing of High Performance Semicrystalline Thermoplastics and Their Composites ,*Solid Freeform Fabr. Symp. Proc.*

[12] Vieille B , Aucher J and Taleb L 2009 Influence of temperature on the behavior of carbon fiber fabrics reinforced PPS laminates, *Mat. Sci. Eng. A.* **517** 51–60.

[13] Senatov F S, Niaza K V, Zadorozhnyy M Y , Maksimkin A V, Kaloshkin S D and Estrin Y Z, 2016 Mechanical properties and shape memory effect of 3D-printed PLA-based porous scaffolds, *J Mech. Behav. Biomed. Mater.* **57** 139–148.

[14] Ning F, Cong W, Qiu J, Wei J and Wang S 2015 Additive manufacturing of carbon fiber reinforced thermoplastic composites using fused deposition modeling, *Compos. B Eng.*, **80** 369–378.

[15] Behzadnasab M and Yousefi A A 2016 Effects of 3D printer nozzle head temperature on the physical and mechanical properties of PLA based product, *12th Int. Semin. Polym. Sci. Technol*, Islamic Azad university,Tehran,Iran.

[16] El Magri A , El Mabrouk K , Vaudreuil S and Ebn Touhami M 2019 Mechanical properties of CF-reinforced PLA parts manufactured by fused deposition modeling, *J. Thermoplast. Compos. Mater.*

IOP Conf. Series: Materials Science and Engineering **783** (2020) 012002 doi:10.1088/1757-899X/783/1/012002

Effect of frequencies on the change of electrical impedance in sand dunes of Ouargla region

N. Remita[1], S. Chihi[1], ML. Mechri[1]

[1] Université de Ouargla, Faculté des Mathématiques et des Sciences de la Matière, Département de Physique, Laboratoire de Rayonnement et Plasmas et Physique des Surfaces (LRPPS), Ouargla 30000, Algeria

Corresponding author: r.naamane@yahoo.fr

Abstract. In this paper, we put the accent on the measurement of real and imaginary parts of electrical impedance and effective loss tangent respectively z', z'' and $tg\delta$ in sand dunes of Ouargla region at a range of frequency intervals between $4,6849\times10^{-2}$ Hz and $1,8593\times10^{+6}$ Hz with AC voltage of 1.000 V using the impedance spectroscopy. The experimental results revealed that the real and imaginary parts of electrical impedance and effective loss tangent have strong frequency dependence. In the frequencies range ($4,6849\times10^{-2}$ – $4,9385\times10^{-1}$) Hz $z'<0$ and z'' change between ($4,8118\times10^{+2}$ – $5,0422\times10^{+1}$) Ω , in the frequencies range ($5,4877\times10^{+1}$ – $2,3426\times10^{+4}$) Hz z' change between 146,04 Ω and 4,8193 Ω and $z''<0$. In high frequencies range ($1,2599\times10^{+5}$ – $1,8593\times 10^{+6}$) Hz z' change between 4,5359 Ω and 5,0567 Ω. The effective loss tangent values increase until they reach a peak, in frequencies range between $4,6849\times10^{-2}$ Hz and $2,5197\times10^{-1}$ Hz, but then the remaining frequency the values decrease until they become constant. The nature of impedance electric variation indicated a possibility to knowledge the polarization of sand dunes of Ouargla region and acquaintance the frequencies when the material is passive devices or active devices. It is also possible to know the quality of equivalent electrical circuit, whether it is induction or capacitance.

1. Introduction

Sand is a granular material where grains diameter varies from 0.02 to 2.0 mm [1, 2]. Its chemical composition is quartz (SiO2) and gypsum (CaSO4.2H2O) [1]

Algerian Sahara has huge quantities of sand dunes. These are among the important natural resources that can be exploited in different fields. Sand is known as a raw material used in industy [3, 4]. Sand is used either in its natural form or after various processes to be ready or use [5, 4]. Sand is an excellent filter for removing sediment and bacteria from water [6, 7] and building [8, 9]. Extracted quartz from sand can be used in glass, optics [3, 10, 11] and to extract silicon which is used in semiconductors and nano-industry.

The previous research works in the field of knowledge of sand dunes in Ouargla region have focused on the diagnostic side [1]. The aim of this work is to measure the electrical impedance of the sand dunes in the terms of frequency, and its effect on the variation of electrical Impedance.

2. Material and methods

2.1. Collection of Sand Samples

The original sample has been taken from the sand dunes of Makhearzha area, located south-west of the city of Ouargla .Ouargla is located in the north of the Algerian Sahara where its geographical coordinates are: latitude is 31◦57' North, longitude 5◦20' East and its mean altitude is 137 m.

Figure.1. Dunes of Ouargla region [1]

2.2. Sample Preparation

The sample of sand was grinded by a mechanical machine containing small pellets of agate, an amorphous type of silica stone. These particles collide with each other by the vibrations caused by the device. The material to be grinded between the pellets is a collision case, which causes crushing. The sample grinding lasted about a quarter of an hour and the sand became very smooth.

2.3. Spectroscopy Techniques

2.3.1. Impedance spectroscopy (IS)

Measuring the complex impedance provides the complex conductivity as a function of frequency. This technique is known as impedance spectroscopy. The electrical impedance connects several physical parameters such as the complex dielectric constant, electrical conductivity and capacity.

The impedance spectroscopy is a powerful proven method of analysis in research and development of materials. This method is relatively simple and non-destructive. The results can be often correlated with a lot of physical properties of materials [12], dielectric properties [13] and the influences of the composition on the electrical conductivity of solids [13, 14]. It is also used as a method for food quality control and pharmaceutical products [15] . Furthermore, it can contribute to the interpretation of fundamental electrochemical processes, usually due to electrochemical phenomena.

This technique is based on the modeling results by equivalent electrical circuits which have the same frequency response as the sample.

The electrical behavior of an electrochemical system can be represented similarly by impedance circuits. Applying a sinusoidal voltage and analyzing the possible system response to obtain information on the various steps involved in the reaction mechanism. All the responses of the system in a very wide frequency range are an impedance diagram.

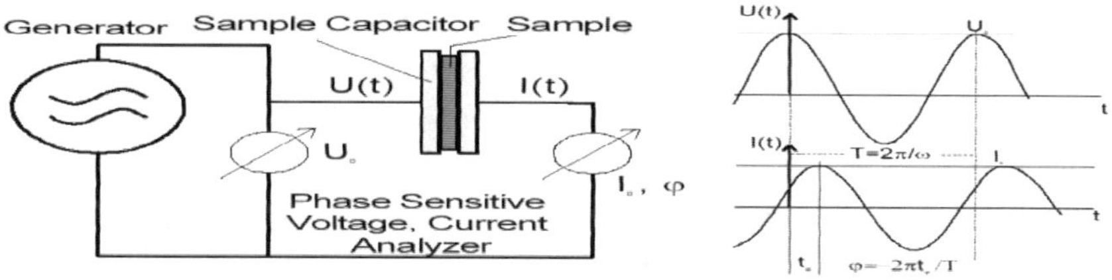

Figure.2. Impedance spectroscopy [16]

2.3.2. Measurement of electrical impedance and effective loss tangent

The complex impedance can be written as follows [17]

$$z_c = z' - iz'' \quad (1)$$

Where

$$z' = \sum_l \frac{R_l}{1+\omega^2 R_l^2 C_l^2} \quad (2)$$

$$z'' = \sum_l \frac{\omega R_l^2 C_l}{1+\omega^2 R_l^2 C_l^2} \quad (3)$$

z' and z'' are the real and imaginary parts of electrical impedance respectively; where ω is the angular frequency, the resistance R_l represents ionic or electronic conduction mechanisms, while the capacitance C_l represents the polarizability of the sample from different components labeled l, which are related to the grain, grain boundary or the electrode interface [17]

If z' is less than zero the material is a passive device, and if it is greater than zero the material is an active device.

When $z''\langle 0$ the equivalent electrical circuit is induction, and when $z''\rangle 0$ the equivalent electrical circuit is capacitance

The effective loss tangent $tg\delta$ can be written as follows [17]

$$tg\delta = -\frac{z'(\omega)}{z''(\omega)} \quad (4)$$

3. Results and discussion

In their previous studies, the researchers like Mahdadi et all [1], focused only on the study of Ouargla sand dune on the formal (morphologies) side, for this reason we will not be able to compare our results with other results in the same field.

The experimental results showed that the real part values of electrical Impedance was negative, in frequencies range between $4{,}6849\times10^{-02}$ Hz and $4{,}9385\times10^{-01}$ Hz, these values indicate that in these frequencies cited early, the Ouargla sand dunes has taken the active devices class (produces power). The real part of electrical impedance takes the greatest positive value at frequency $5{,}4877\times10^{+01}$ Hz, but its value begins to decrease until it stabilizes at all remaining frequencies (Figure3(a)). This change in the real part of electrical impedance values shows that Ouargla region sand dune plays the role of passive devices (consuming energy) in this case.

But the imaginary parts take high positive values whereas the real part is negative which indicate that the equivalent electrical circuit is induction.

In the frequencies range between $5{,}4877\times10^{+01}$ and $2{,}3426\times10^{+04}$ the imaginary parts values are negative. In this case the equivalent electrical circuit is capacitance.

In the other frequencies, the imaginary parts increase to reach positive values and finally stabilize (Figure.3. (b)).

Between frequencies $4{,}6849\times10^{-02}$ Hz and $2{,}5197\times10^{-01}$ Hz, The effective loss tangent values increase until they reach the peak, then they decrease and remain constant (Figure.3. (c)).

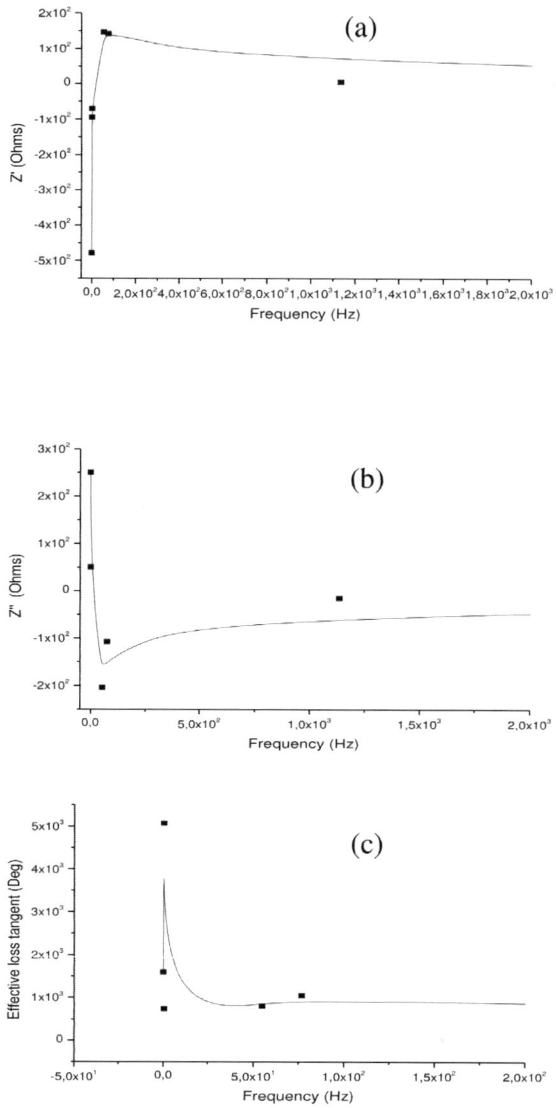

Figure.3. (a) Real parts of electrical impedance and Frequency,
(b) Imaginary parts of electrical impedance &Frequency,
(c) Effective loss tangent &Frequency

4. Conclusion

In the present work, we have mainly reported the electrical impedance, effective loss tangent, and calculate their values in sand dune of Ouargla region, using the Impedance spectroscopy method. The experimental results showed a correlation between real, imaginary parts of electrical impedance and frequency. At low frequencies the values of real parts were negative (The Ouargla sand dunes has been taken the active device class), while the imaginary parts have taken high positive values (The equivalent electrical circuit is inductive). At medium and high frequencies real parts were positive (Ouargla region sand dune plays the role of passive devices) but imaginary parts took negative values (The equivalent electrical circuit is capacitive).

References

[1] Mahdadi N , Chihi S , Bouguettaia H , Beddiaf S , Mechri ML,(2017) Chromatic Classification of Ouargla (Algeria) Dunes Sand: Determination of Main Compositions and Color Causes, by Using XRD, FTIR and XRF, *Springer Science+Business Media Dordrecht,silicon* **9**, 211-221

[2] Marc P, Jacques G (2006) Handbook of soil analysis: mineralogical, organic and inorganic methods. *Springer, Berlin*, p 16

[3] Platiasa S, Vatalisa KI, Charalampidesa G (2014) Suitability of quartz sands for different industrial applications. *Procedia Econ Finance* **14**:491–498

[4] Hashim U, Ehsan AA, Ahmad I (2007) High purity polycrystalline silicon growth and characterization. *Chiang Mai J Sci* **34(1)**:47–53

[5] Demyanova LP, Rimkevich VS, Buynovskiy AS (2011) Elaboration of nanometric amorphous silica from quartz-based minerals using the fluorination method. *J Fluor Chem* **132**:1067–1071

[6] Haig SJ, Quince C, Davies RL, Dorea CC, Collins G (2014) Replicating the microbial community and water quality performance of full-scale slow sand filters in laboratory-scale filters. *Water Res* **61(1)**:141–151

[7] Bendida A, Tidjani AE, Badri A, Kendouci MA, Nabou M (2013) Treatment of domestic wastewater from the town of Bechar by a sand filter (sand of Beni Abbes Bechar Algeria). *Energy Proc* **36**:825–833

[8] Leghrieb Y, Mitiche R, Bentebba MT, Djouhri M, Kriker A (2012) The manufacture of raw brick from the saharan Sand-Based Mortar of Ouargla (Located in the Septentrional Sahara. Algeria) for use in Arid Regions. *Arab J Sci Eng* **37**:2149–2161

[9] Guettala S, Mezghiche B, Mellas M (2010) Adding finely crushed dune sand to cement on the evolution of hydration of pasta Portland cement. *Asian J Civ Eng (building and housing)* **11(2)**:241–251

[10] Hajjaji W, Jeridi K, Seabra P, Rocha F, Labrincha JA, Jamoussi F (2009) Composition and properties of glass obtained from Early Cretaceous Sidi Aich sands (central Tunisia). *Ceram Int* **35**:3229–3234

[11] Saikia BJ, Parthasarathy G, Sarmah NC, Baruah GD (2008) Fourier–transform infrared spectroscopic characterization of naturally occurring glassy fulgurites. *Bull Mater Sci* **31(2)**:155– 158

[12] Ding, SJ. Chang, BW, Wu, CC., Lai, MF, and Chang, HC, (2005). Impedance spectral studies of self-assembly of alkanethiols with different chain lengths using different immobilization strategies on Au electrodes. *Analytica Chimica Acta* **554** Issues (1-2), 43.

[13] Young IJ, (2000) .Dielectric measurement and analysis for the design of conductor/insulator artificial dielectrics. *IEE Pros –Sci* . **147** N°4, 4773.

[14] Zekri L, Zekri N and Clerc JP, (2004). Modelling of impedance spectroscopy of composites by electrical networks. *Journal of Physics* . **37**, 535.

[15] Hong T, Chen Y, Sun H and Jepson W.P, (2001). Monitoring corrosion in multiphase pipelines *Materials and Corrosion* **52**, 590.

[16] Sylvestre A, (2006). *Apport de la spectroscopie diélectrique basse fréquence dans l'analyse de matériaux isolants à forte permittivité*, Energie électrique, Université Joseph-Fourier Grenoble I,

[17] Mahmoudi Chenari H, Golzan M.M , Sedghi H, Hassanzadeh A, Talebian M, (2011). Frequency dependence of dielectric properties and electrical conductivity of Cu/nano-SnO2 thick film/Cu arrangement. *Current Applied Physics.* **11**, 1071-1076.

Elaboration, Thermal and Structural Approach of Manganese Phosphate Glasses Inside the System of K_2O-MnO_2-P_2O_5

M. Jerroudi[1,*], S.Yousfi[2], L. Bih[2], L.Bejjit[2], M. Haddad[3], B. Manoun[4], and P. Lazor[5]

[1]Equipe de Physico-Chimie de la Matière Condensée, PCMC, Faculté des Sciences de Meknès, Maroc.
[2]Département matériaux et procédés, ENSAM Meknès, Université Moulay Ismail, Maroc.
[3]Laboratoire de Spectrométrie, des Matériaux et Archéomatériaux (LASMAR), Faculté des Sciences de Meknès,Université Moulay Ismail, Maroc.
[4]Laboratoire des Sciences des Matériaux, des Milieux et de la Modélisation (LS3M), Univ Hassan 1er, 26000, Khouribga, Maroc.
[5] Department of Earth Sciences, Uppsala University, SE-752 36, Uppsala, Sweden.
Corresponding e-mail address: meryjerroudi1994@gmail.com

Abstract. Glasses of the composition (50-x/2) K_2O-xMnO_2-(50-x/2) P_2O_5 with ($0 \leq x \leq 30$%mol), have been prepared by standard melt quenching procedures. The amorphous state of the glasses is evidenced using the X-Ray diffraction. Their physical properties were characterized by thermal analysis and density measurements. The density and the glass transition temperature increase with increasing in MnO_2 content. To study the structural role of MnO_2 oxide in studied glasses, the FTIR and Raman spectroscopies have been employed. It was highlighted that the presence of MnO_2 allows the depolymerisation of the phosphate chains and formation of covalent bonds P-O-Mn, which replace P-O-P and P=O linkages.

Keywords: Manganese glasses, Glass transition temperature, density, Infrared spectroscopy, Raman spectroscopy.

1. Introduction

This article will introduce manganese phosphates glasses as consider among the glasses containing transitions metal ions (TMI), which have drawn much attention, because of their electric, optical and magnetic properties. These properties come from the presence of TMI in different multivalent states [1]. The most interesting glass structure modifier is manganese; it was used in metallurgy, as a component and alloy addition increasing their hardness and resistance to abrasion [2]. It was reported in the literature, that the manganese ions exist in different valence states with different coordination in glasses matrices [3, 4]. The content of manganese in different coordination in different valence state is related to the quantitative properties of modifiers and glasses formers, size of the ions in the glass, and the mobility of the modifier cation…[4]. Due to their interesting properties and their coordination, the glasses containing manganese have been subject of several publications [5,1,6].

In the present work, the structure of glasses with the general composition of (50-x/2) K_2O-xMnO_2-(50-x/2)P_2O_5 ($0 \leq x \leq 30$ % mol) is investigated using FTIR and Raman spectroscopies to elucidate the

IOP Conf. Series: Materials Science and Engineering **783** (2020) 012003 doi:10.1088/1757-899X/783/1/012003

structural change as a function of composition. Other physical properties such as density, molar volume and glass transition temperature were also studied with the composition.

2. Experimental procedure

The phosphate glasses of composition (50-x/2) K_2O-xMnO_2-(50-x/2)P_2O_5; x=0,10,20 and 30% mol of MnO_2, were prepared by melt-quench process. The $NH_4H_2PO_4$, K_2CO_3, and MnO_2 are used as raw materials. Each composition was carefully weighted and then mixed in a porcelain crucible placed in the electrical furnace. The mixture was initially heated at 300°C for 12 hours in order to remove NH_3 and H_2O. Then, the temperature increases at 600°C in order to remove CO_2 gas in K_2CO_3. The batch was finally quenched to room temperature under air in order to produce vitreous structure. The obtained glasses were homogeneous and colored. The solids were kept in desiccator to prevent possible moisture.

3. Results and discussions

3.1. Glassy formation and XRD analysis

Fig.1 displays the ternary diagram which report the all formulations investigated in this work. The X-ray diffractograms of the (50-x/2)K_2O-xMnO_2-(50-x/2)P_2O_5 glasses are shown in **Fig.2**. No peak of crystallization was observed, which indicates the amorphous state of all the elaborated materials.

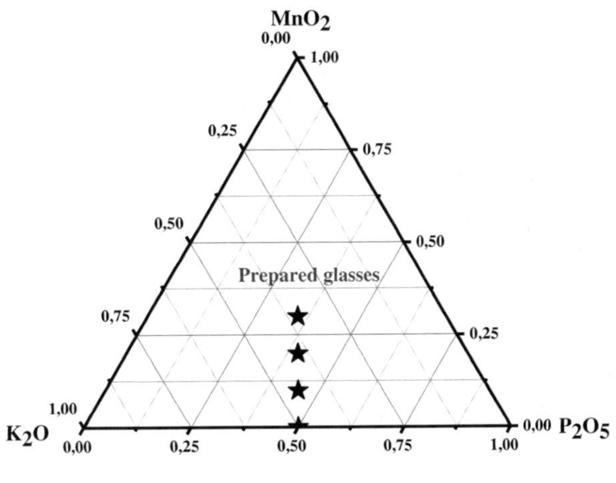

Figure 1. Location of studied glasses in the system (50-x/2) K_2O-xMnO_2-(50-x/2) P_2O_5 in the ternary diagram.

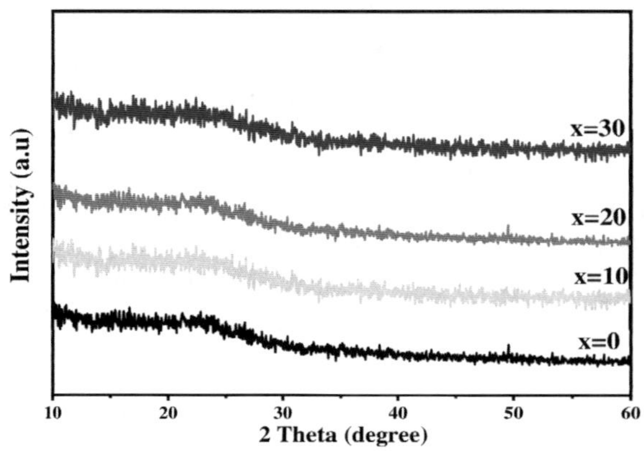

Figure 2. X-ray patterns of the studied glasses

3.2. Density and molar volume

The density values for different bulk of samples are represented in **Fig.3**. For the studied glasses, (50-x/2)K_2O-(50-x/2)P_2O_5–xMnO_2, we observe that the density values increase from 2.35 for x=0 to 2.76 for x=30 %mol in MnO_2 content. This can be proved by the formation of P-O-Mn covalent bonds linkage which reticule the phosphate network. However, the molar volume decreases from 50.24 for x=0 to 39.39 cm^3/mol for x=30 %mol in MnO_2 content. This variation can be due to the decrease of the interstitial space between the polyhedra. Therefore, the planned structure is closed. One can suppose that the substitution of KPO_3 by MnO_2 induces the conversion of some bridging oxygen (BO) to non-bridging oxygen (NBO). Similar variation of the density and molar volume have also been observed in systems contain manganese oxide, in 20K_2O-xMnO-(80-x)P_2O_5 [7], in(50-x)K_2O-xMnO-50P_2O_5 [6], and (50-x/2)Na_2O-xMnO-(50-x/2)P_2O_5 [1].

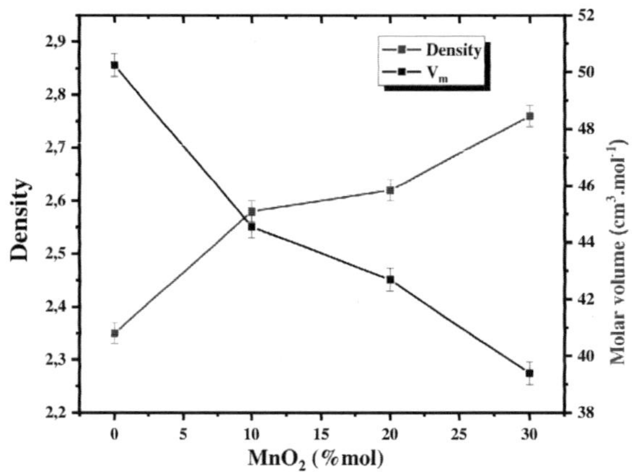

Figure 3. Variation of the density and molar volume of glass with manganese concentration

3.3. DSC study

The DSC curves of the studied glasses are shown in **Fig.4**. The evolution of glass transition temperature T_g as a function of MnO_2 amount is represented in **Fig.5**. An increase in T_g from 214°C for x=0 to 372°C for x=30 %mol MnO_2 was seen in prepared glasses. The increase of T_g indicates an increase of the network crosslink strength of the structure as Mn^{2+} ions are introduced to the glassy-matrix. It is worth to mention, that the nature of bonding is responsible for this variation in T_g. For instance, the K_2O rich glasses are predominantly ionic and become covalent with the addition of MnO_2 in KPO_3 matrix, leading to the formation of P-O-Mn covalent bonds. This result is explained by the electrical field strength which is larger for Mn^{2+} than that of K^+. The same change of T_g with manganese content was also observed in the system $20K_2O$-xMnO-$(80-x)P_2O_5$ [6] and the (50-x)K_2O-xMnO-$50P_2O_5$ system [7].

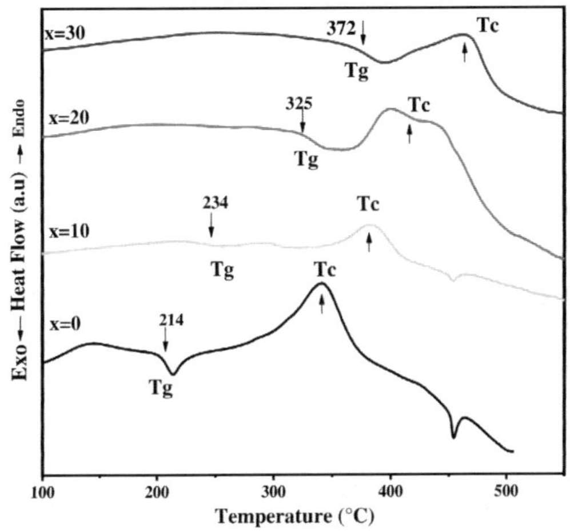

Figure 4. Differential scanning calorimetry (DSC) curves for glass samples

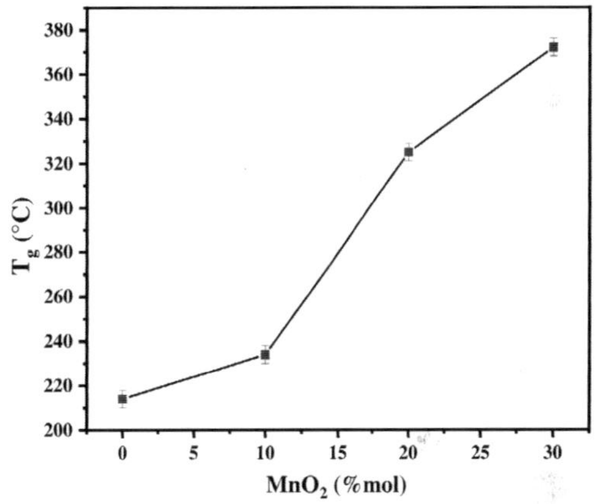

Figure 5. Composition dependence of glass transition temperature for the studied glasses.

3.4. FTIR and RAMAN spectroscopy

FTIR and Raman spectra of $(50-x/2)K_2O$-$xMnO_2$-$(50-x/2)P_2O_5$ glasses with various concentration of manganese oxide are presented in **Fig.6** and **Fig.7** respectively. The assignment of different bands is realized based to the comparison of our result with the literature data [1,8].

FTIR spectra (**Fig.6**), reveal the disappearance of the $\upsilon_{as}(P=O)/\upsilon_{as}(PO_2)$ band at 1274 cm^{-1} as the concentration of MnO_2 increases. This statement was be explained by the conversion of some P=O bonds into P-O-Mn bonds. A shoulder at about 1031 cm^{-1} is an indicator of the presence of some pyrophosphates unities. These structural groups are also evidenced from Raman analysis.
From the Raman spectra (**Fig.7**), we notice the appearance of a new band at 1020 cm^{-1} attributed to pyrophosphate groups (Q^1). Therefore, we can state that the MnO_2 causes a depolymerisation of the glassy matrix; its presence induces the conversion of some metaphosphate structural units (Q^2) to pyrophosphate ones (Q^1).

IOP Conf. Series: Materials Science and Engineering **783** (2020) 012003 doi:10.1088/1757-899X/783/1/012003

The introduction of MnO_2 causes progressive spectral changes. In particular, the stretching mode band of P-O-P shifted to higher frequencies from 689 cm^{-1} in KPO_3 to 710 cm^{-1} for $35K_2O$-$30MnO_2$-$35P_2O_5$ glass. This result can be interpreted in term of the increase of the covalent character of P-O-P bonds as Mn^{2+} ions replace the K^+ ions. Since the potassium belongs to column one of the periodic table, it has a strong ionic character compared to Mn.

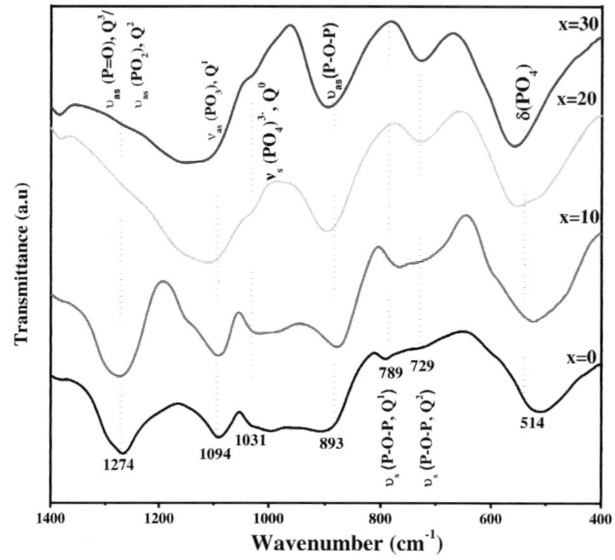

Figure 6. Infrared spectra of the studied glasses

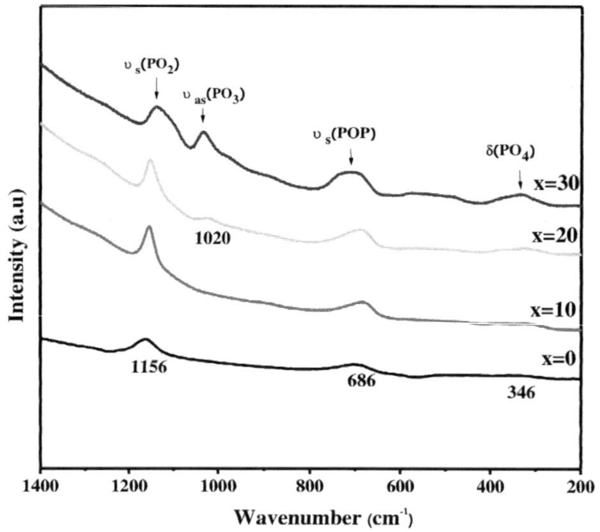

Figure 7. Raman Spectra of the studied glasses.

4. Conclusion

In this work, we have investigated phosphate glasses belonging to the ternary K_2O-MnO_2-P_2O_5 system. The amorphous state of the samples was confirmed by X-Ray diffraction. From the physical properties results, it has been shown that the density and T_g increase, which indicate that MnO_2 reticule the phosphate network by the formation of P-O-Mn bonds. Infrared and Raman spectroscopies allowed us to follow the evolution of glass structure as the introduction of MnO_2 content. One can notice the modification of the vitreous network when MnO_2 oxide is introduced suggesting the shortening of the infinite metaphosphate chains.

Acknowledgements

The authors would like to thank the CNRST (Morocco) and OCP foundation for their financial support of this work in the framework of PPR & Apphos projects.

References

[1] Omrani R, Krimi S,Videau J J, Khattech I, El Jazouli A, and Jemal M (2014), *J. Non. Cryst. Solids*, **389**, pp. 66–71.

[2] Szumera M and Wacławska I (2012), *J. Therm. Anal. Calorim*, **108**, no. 2, pp. 583–588.

[3] Pascuta P, Bosca M, Borodi G, and Culea E (2011), *J. Alloys Compd*, **509**, no. 11, pp. 4314–4319.

[4] Reddy M S, Krisha G M, and Veeraiah N (2006), *J. Phys. Chem. Solids*, **67**, no. 4, pp. 789–795.

[5] Ardelean I, Cozar O, Filip S, Pop V, and Cenan I (1996), *Solid State Commun*, **100**, no. 8, pp. 609–613,

[6] Ahmina W, El Moudane M, Zriouil M, and Taibi M (2016), *Phase Transitions*, **89**, no. 11, pp. 1051–1061,.

[7] Ahmina W, El Moudane M, Zriouil M, and Taibi M (2017), *J. Mater. Environ. Sci.*, **7**, no. 3, pp. 694–699.

[8] Jerroudi M, Bih L, Azrour M, Manoun B, Saadoune I, and Lazor P (2019), *J. Inorg. Organomet. Polym. Mater.* 10.1007/s10904-019-01213-0

Swelling Behavior And Mechanical Properties In Filled Elastomeric Nano-Composites In Contact With Organic Solvents

A Mdarhri[1], F El Haouzi[1], C Brosseau[2], I El Aboudi[1], M Bentoumi[3,4] and M EL Azhari[1]

[1]Laboratory of Condensed Matter and Nanostrucres (LCMN), Faculty of Sciences & Techniques -Cadi Ayyad University, A. Khattabi Marrakech BP 549, 40 000, Morocco.
[2]Lab-STICC, Université de Brest, UMR CNRS 6285, CS, 93837, 6 avenue Le Gorgeu, 29238, Brest Cedex 3,France
[3]Laboratory of applied optics, Institute of Optics and Precision Mechanics, University Ferhat Abbas, Sétif-1, Algeria 19000.
[4]Department of Civil Engineering University Bordj Bou Arréridj, Algeria 34000.

E-mail: a.mdarhri@uca.ma

Abstract. The effect of organic solvents on the transport process and mechanical properties in elastomeric nanocomposite materials is reported. The investigated samples are formed by a semi-crystalline ethylene-co-butyl acrylate (EBA) polymer filled with hard spherical carbon black (CB) nanoparticles. The swelling behavior was studied at room temperature by immersion the dried samples in two selected solvents, i.e. toluene and xylene. The transport of the molecules of a given solvent in these filled elastomeric composites is found to follow a Fickian diffusion mechanism. Mechanically speaking, the stress-strain curves of uniaxial tensile tests pre- and post- swelling highlight a remarkably decrease of the Young's modulus and strength of the swollen samples. This behavior can be attributed to the decrease of the load transfer density between the matrix and the CB in the presence of the solvent. The Mooney-Rivlin model is used to capture the physics of the stress-strain curves in both dry and swollen samples. The results reported in this experimental investigation can be useful for some demanding applications.

1. Introduction

Carbon black (CB) filled elastomers are among the most commonly studied composite materials due to their favorable combination of remarkable mechanical and electrical properties [1]. It is by now well established that these properties are under the dependence of the particle shape, particle aggregation structure, porosity, polymer rheology, and processing conditions [2]. In practice point of view, the aggressive environmental factors including chemical attack [3], [4], heat or stress can deteriorate the physical and mechanical properties of these materials systems such as stiffness, tensile strength, and toughness [5]. Particularly, the contact with organic solvents can generate the change in the mass or volume expansion (dimensional instability) of the

material due to a possible penetration of solvent [6]. This phenomenon, namely swelling [7], is due to solvent transport inside filled elastomeric composites and it depends on the filler dispersion state in the polymer matrix, nature of filler and physicochemical properties of a given solvent [8].
The aim of this work is firstly to understand the swelling phenomenon in Ethylene-Butyl Acrylate (EBA)/CB composites by using two selected commercial solvents that are toluene and xylene. The effect of the CB particles in the kinetics absorption of each solvent is also examined according to the Fick's law. Secondly, the mechanical behavior of swollen and unswollen materials obtained from tensile tests is investigated at room temperature. The stress-strain curves are then correlated to kinetics absorption results and interpreted according to the semi-empirical approach proposed by Monney-Rivlin.

2. Experimental

2.1. Materials

Ethylene Butyl Acrylate copolymers (EBA) filled by acetylene CB (Denka Black) were studied. These composites were obtained from Borealis AB (Sweden)[9]. The density of the CB particles is 1.8 $g.cm^{-3}$ and its specific surface area is 63 $m^2.g^{-1}$. The glass temperature transition of the neat polymer matrix is $T_g = 198\,K$, and its crystallinity ratio is ~20%. The CB filled polymer samples were fabricated by mechanical mixing as is detailed elsewhere[9], [10]. For this study, a series of the composite samples was used containing three CB volume concentrations. Figure 1 illustrates a typical Scanning Electronic Microscopy (SEM) picture indicating a good dispersion of CB nanoparticles in the EBA matrix.

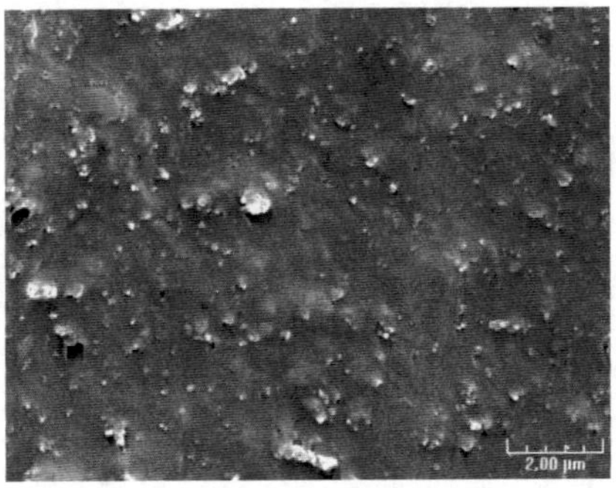

Figure 1. A typical SEM images of fractured surface of CB/EBA samples with 8.42 vol.% of CB.

2.2. Swelling test

This study deals with the investigation of the transport phenomenon in a series of carbon black/elastomeric composite samples under swelling tests and its effect on the mechanical properties. Two chemical compounds methyl derivatives of aromatic hydrocarbons of benzene but with increasing size and volume, i.e. toluene and xylene are used. The selected solvents allow us to study the swelling kinetics without destroying or dissolve the composite materials under test. The swelling behavior of the nanocomposite was studied by immersion the dried samples in a pure solvent at room temperature until saturation. At specific times, the samples were taken out from the solvents, the excess of solvent was removed from their surfaces using filter paper, and then, samples are weighted. The percentage of mass change (Q) was calculated using the following relation [11]:

$$(Q\%) = \frac{M_2 - M_1}{M_1} \times 100 \quad (1)$$

where M_1 and M_2 are the mass of sample before and after immersion respectively.

2.3. Mechanical testing

To gain insight on the effect of swelling on the mechanical response of filled elastomers, uniaxial tensile tests on dry and swollen samples were carried out using Instron 3396 testing machine operated at room temperature. For all samples, tests were conducted at a constant strain rate, i.e. $2mm.min^{-1}$. To ensure repeatability of the results, at least three samples were used in each test.

3. Analysis and Modeling

The materials swelling kinetics in both solvents is analyzed according to the diffusion Fick's law expressed by the equation [7], [12]:

$$\Delta m_t / \Delta m_\infty = 4(Dt / \pi e^2)^{1/2} \tag{2}$$

where, Δm_t and Δm_∞ are defined as change in mass at time t relative to $t = \infty$ for Δm_∞ (saturation), and e and D denote the thickness of the sample and the coefficient of Fickian diffusion respectively. Mechanism of transport can be analyzed by the power law according to the following equation [13]:

$$Q_t/Q_\infty = kt^n \tag{3}$$

where Q_t is the percent solvent up taken, the constant n gives the mode of sorption mechanism. If $n = 0.5$ mode of transport is Fickian, n=1 indicates non- Fickian mode of transport while n between 0.5 and 1 indicates anomalous behavior

The experimental stress–strain curves can explained within the Mooney–Rivlin empirical model[14] which is often used by rubber scientists [15]. For the uniform extension with uniaxial tension the stress σ is related to the extension ratio λ (which is $1 + \varepsilon$, where ε is strain) by:

$$\sigma = 2\left(C_1 + \frac{C_2}{\lambda}\right)\left(\lambda - \frac{1}{\lambda^2}\right) \tag{4}$$

where C_1 and C_2 are material constants. The fitted line of the normalized stress $\sigma_n = \sigma/2(\lambda - 1/\lambda^2)$ versus λ^{-1} is straight with constant slope C_2 and the intercept at $\lambda^{-1} = 0$ provides the value of C_1.

4. Results and Discussion

4.1. Absorption kinetics

Figures 2a and 2b show the percentage change in mass Q versus time t for elastomeric matrix filled with three values of CB volume fractions. All curves highlight the same trend characterized by a rapidly increase of Q followed by a saturation level. The magnitude of the mass change compared to the toluene where the gap between the two solvents increases from 10.25% to 18.38% when the volume fraction of CB particles increases from 8.42 to 19.9 vol%. In despite of its higher molecular weight [16], the xylene penetrates easily in the composite samples compared to the toluene. From the figure 2a, the equilibrium sorption decreases with increase of carbon black amount. This behavior can be attributed to the presence of the three-dimensional network formed by the CB clusters. These clusters which exist on finite or infinite paths above the percolation threshold [17], act as barriers to movement of the toluene and their density increases when CB volume fraction increases. The results obtained in cases of the samples immersed in the xylene (figure 2b) show a similar effect of the CB particles on the equilibrium sorption.

It is found that the materials swelling kinetics in both solvents follow the diffusion Fick's law. The values of the effective diffusion coefficient (D) of the composites were deduced by plotting $\Delta m_t / \Delta m_\infty$ versus $t^{\frac{1}{2}}/e$ in the linear region with the slope is $4(D\pi)^{1/2}$[18]. The obtained values are ranged from 1.96 10^{-7} to 3.47 10^{-7} and from 3.01 10^{-7} to 5.47 10^{-7} $cm^2.s^{-1}$ for toluene and xylene respectively.

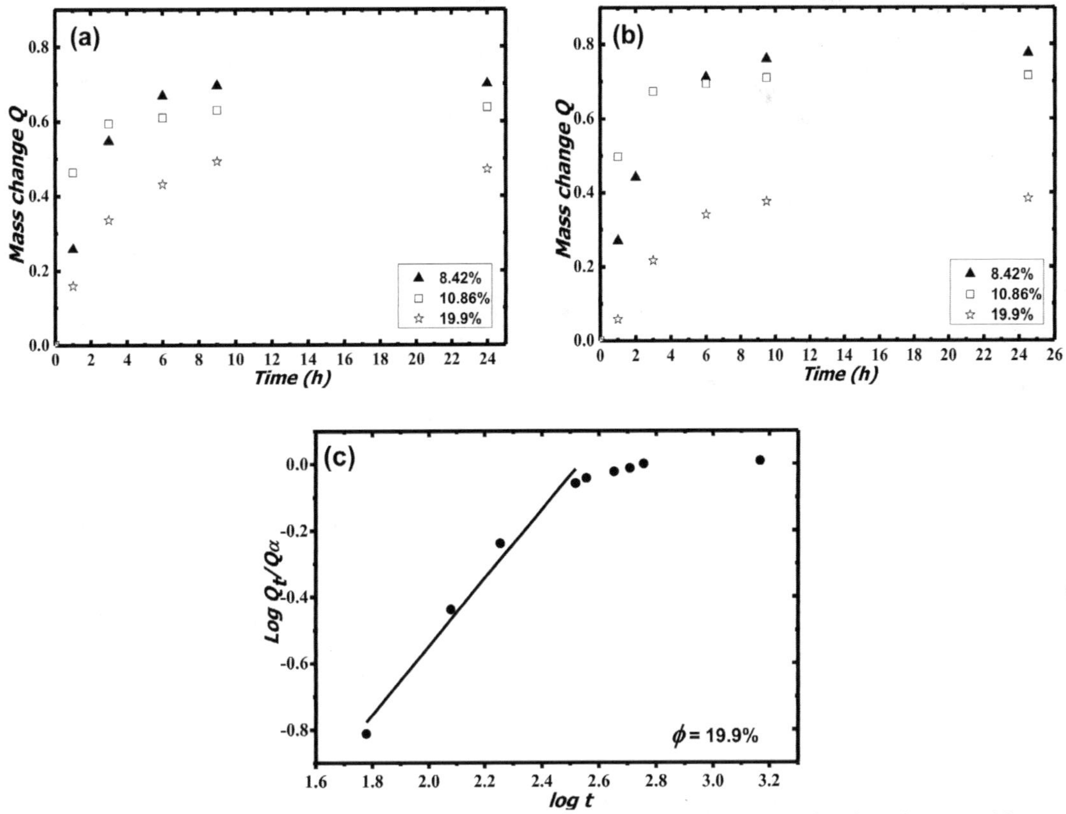

Figure 2. (a) Change in mass versus time for samples with three CB volume fractions immersed in toluene.(b) Same as in (a) for xylene. (c) Plot of log Q_t/Q_∞ versus log t for diffusion of the swollen (xylene) sample with 19.9 vol.% of CB. The solid line is a fit to the data according to $Q_t/Q_\infty \alpha t^n$ with n =1.03.

Figure 2c illustrates a typical linear relationship between log Q_t/Q_∞ and $logt$ before saturation level. The obtained values of n for the swollen sample by toluene and xylene solvents are reported in the Table 1. It follows that, for both solvents, two modes are responsible of the kinetics data: (i) the first one namely anomalous diffusion is found for composites with 8.42 and 19.9 vol.% of CB, owing to the coupling of Fickian and non- Fickian behaviors and (ii) the second called less Fickian for the samples with10.86 vol.% of CB.

Table 1. The calculated values of the exponent n according to the power law $Q_t/Q_\infty \alpha\, t^n$ as function of CB content.

	Fit exponent n for swollen samples	
Vol.% of CB	**in toluene** [18]	**in xylene**
8.42	0.64	0.56
10.86	0.36	0.35
19.9	0.89	1.03

4.2. Tensile behavior

The stress–strain curves of composite samples with different CB content are reported in figure 3. Firstly, the curves of unswollen samples highlight large stress-strain deformations where the largest deformation reaches ~800% (not shown) [19]. The addition of the CB nanoparticles improves significantly the mechanical properties of the dry elastomeric matrix such as stiffness and tensile strength indicating a high matrix-carbon black interfacial density due to a good dispersion of CB in matrix [8]. When the solvent molecules penetrate in the composite samples, a remarkable decrease in their mechanical characteristics is observed in figure 3. The mechanical properties of samples exposed to xylene highlight a higher reduction compared to swollen samples in a toluene solvent that is in agreement with the absorption kinetics results presented above. The significant reduction of the properties mechanical (Yong's modulus, yield stress, ...) in the swollen samples may ascribed to a formation of a liquid by the solvent's molecules from the surface to the bulk of composite samples [20, 21]. This liquid forms an interface which would works like a preferential area affecting elastomeric/CB interface causing a bad stress transfer between the charge and the matrix. This mechanism has been demonstrated on methanol in PMMA and dodecane in PS system [22] and on epoxy/acetic acid-water system [23].

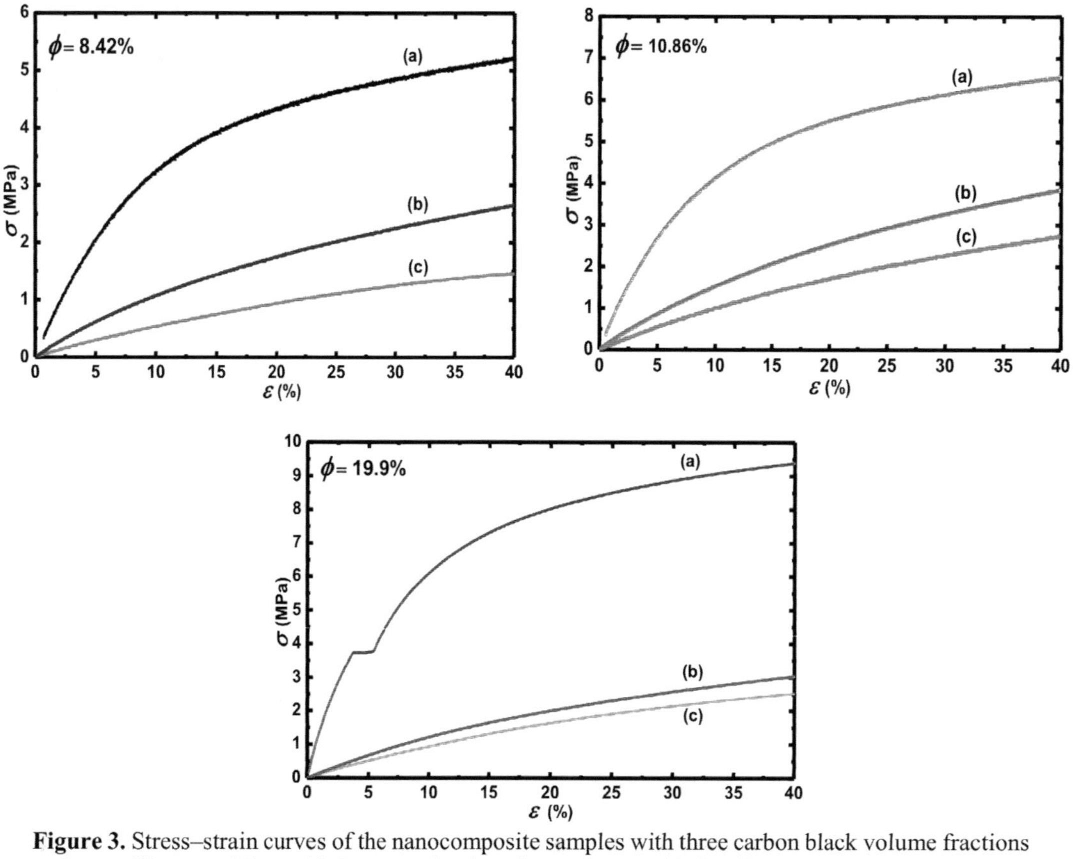

Figure 3. Stress–strain curves of the nanocomposite samples with three carbon black volume fractions under swelling conditions. (a) Dry sample plotted as reference (b) Swollen sample by using toluene, (c) Swollen sample by using xylene.

Table 2. The values of the Mooney-Rivlin constants $C_1$and $C_2$of EBA/CB samples before and after swelling.

Vol.% of CB		**8.42**			**10.9**			**19.9**	
Monney-Rivlin constants	**Unswollen Sample**	**Swollen by toluene**	**Swollen by xylene**	**Unswollen Sample**	**Swollen by toluene**	**Swollen by xylene**	**Unswollen Sample**	**Swollen by toluene**	**Swollen by xylene**
C_1**(MPa)**	0.527	0.474	0.315	0.497	0.436	0.689	0,646	0.256	0.574
C_2**(MPa)**	2.567	1.434	0.731	3.511	2.41	1.178	4.884	2.031	2.003

5. Conclusion

This work deals with swelling behavior and mechanical properties of dry and swollen composite samples at room temperature. The material is made by a homogenous dispersion of carbon black nanoparticles in an elastomeric matrix. Firstly, the time dependence of the transport of solvent molecules, i.e. toluene and xylene, in these heterogeneous systems is found to follow a Fickian diffusion mechanism. The CB amount and the nature of the solvent dependence of the absorption curves are related to some microstructural characteristics of the composite samples. Mechanically speaking, the swollen samples highlight a remarkably reduction in their mechanical responses compared to dry samples. The effect of swelling on the measured properties is mainly linked to a preferential migration of solvent at filler-rubber interface altering load transfer and correlated to the absorption results. The Mooney-Rivlin model is used to capture the physics of the stress-strain curves in both dry and swollen samples.

Acknowledgments
This work was partially supported by the Centre National de la Recherche Scientifique et Technique -CNRST, Morocco.

References

[1] Donnet J B Bansal R C Wang M J 1993 *Carbon black: science and technology*, New York
[2] Bokobza L 2013 *Rubber Chem. Technol.* **86** 423.
[3] Tanaka T Ishiwata S Ishimoto C 1977 *Phys. Rev. Lett.* **38** 771.
[4] Sato Matsuo E Tanaka T 1988 *J. Chem. Phys.* **89** 1695.
[5] Sombatsompop N 1998 *Polym. Plast. Technol. Eng.* **37** 19.
[6] Abu-Abdeen M Elamer I 2010 *Mater. Des* **31** 808.
[7] Crank J 1975 *The mathematics of diffusion*, 2d ed. Oxford, Clarendon Press.
[8] Ekebafe O L M.Ekebafe O Ugbesia O S Ewansiha J C 2011 *Sci. and Math.* 127.
[9] Mdarhri A Brosseau C Zaghrioui M El Aboudi I 2012 *J. Appl. Phys.* **112** 034118.
[10] Brosseau C Mdarhri A Vidal A 2008 *J. Appl. Phys.* **104** 074105.
[11] Trakarnpruk W Porntangjitlikit S 2008 *Renew. Energy* **33** 1558.
[12] Boulic F 2000 *Contribution to the study of relations between mesostructured network elasticity and macroscopic properties of filled polymers*,Thesis, Universitéde Bretagne Occidentale, France
[13] Abraham J J.Maria H George S C Kalarikkal N Thomas S 2015 *Phys. Chem. Chem. Phys.* **17** 11217.
[14] Han W H Horka F Mckenna G B 1999 *Math. Mech. Solids* **4** 139.
[15] Van Krevelen D W and TeNijenhuis K 1990 *Properties of Polymers Their Correlation with Chemical Structure; their Numerical Estimation and Prediction from Additive Group Contributions. Amsterdam,* Nederlands: Elsevier Science & Technology Books
[16] Seehra M S Yalamanchi M Singh V 2012 *Polym. Test.* **31** 564.
[17] Mdarhri A Brosseau C Carmona F 2007 *J. Appl. Phys.* **101** 084111
[18] Elhaouzi F Mdarhri A Brosseau C El Aboudi I Almaggoussi A 2019 *Polym. Bull.* **76** 2765.
[19] Elhaouzi F Nourdine A Brosseau C Mdarhri A El Aboudi I Zaghrioui M *2019 Polym. Compos.* ***40*** *3005.*
[20] Vesely D 2011 *Polymer* **42** 4417.
[21] Brosseau C Bourbigot C Beroual A Queffelec P Le Mest Y Loaec J 1997 *J. Appl. Phys.* **81** 882.
[22] Morrissey P Vesely D 1999 *Polymer.* **41** 1865.
[23] Dutheillet Y Mantle M Vesely D Gladden L 1999 *J. Polym. Sci. Part B Polym. Phys.* **37** 3328.
[24] Piazza R 2012 *Phys. Today.* **65** 56.

IOP Conf. Series: Materials Science and Engineering **783** (2020) 012005 doi:10.1088/1757-899X/783/1/012005

Crystal structure: How to use data quality Indicators to select a high resolution cutoff

M.AIT MOUHA[*,1]; D.TLAMSAMANI[1] ; K.YAMNI[1]

[1] *EMaMePS, Department of Chemistry, Faculty of Sciences, Moulay Ismail University, Meknes ,Morocco.*

* *Corresponding author: Tel: +212674031273, Email:aitmouha.m@gmail.com*

Abstract: To get a good quality crystal structure, we should first do the initial analysis of the diffraction data. There are many global indicators of the quality of the diffraction data. Accentuated in this paper are the $\mathbf{d_{min}}$, $\mathbf{d_{eff}}$,**C**, $\mathbf{R_{merge}}$ and $\mathbf{R_{r.i.m.}}$ Based on the results obtained, this paper will overview in our understanding of how to use these indicators to select a high resolution cutoff that leads to the best crystal structure, using practical examples.

1.Notation

d_{min} **:** the nominal resolution
$\mathbf{d_{eff}}$: the effective resolution
$\mathbf{R_{merge}}$**:** the merging factor
$\mathbf{R_{r.i.m}}$: the redundancy independent merging R factor
I/σ(I): Signal-to-noise ratio
C: the completeness
$\mathbf{N_{refl}}$: the number of reflections
N: Redundancy.
CC*: the correlation coefficients
Obs/param: N of reflections used/N of lsq parameters
wR: The weighted R-factor based on F^2
R: The unweighted residual factor based on F
S: is the goodness of fit

2.Introduction

A series of operations are carried out on the data, which is normally referred to as 'data reduction' following integration of the spots on a set of X-ray diffraction images to produce a list of reflection intensities. Many papers (e.g. [1] [2]) have well documented algorithms and methods for data reduction. The findings will not be reiterated in this paper. We can analyze the quality of diffraction data when we have a set of scale factors to put all the intensities on a common scale and improved estimates of the error on the intensity.

The importance of confirming that the highest quality data set of a given crystal is obtained cannot be overemphasized, particularly in light of the fact that the diffraction experiment is the final real experimental step in the course of a structure determination. Many papers and comparative studies on the evaluation of the quality of the shaping diffraction data have been published [3] [4] [5]. They comparatively examine the evaluation of the quality of crystal structure.

This paper will give and describe an outline of the global indicators of the quality of diffraction data that have been described to date will be given. By using practical examples, this paper will also study their strengths as well as their usefulness to select a high-resolution cutoff.

IOP Conf. Series: Materials Science and Engineering **783** (2020) 012005 doi:10.1088/1757-899X/783/1/012005

3. Indicators of data quality

Taking for granted that the data reduction and the data collection strategy were carried out in a significant and practical way.
Several parameters, such as $d_{min,}$ C, R_{merge}, $R_{r.i.m}$ and I/σ(I) can be used to probe the internal data quality.

3-1 The resolution of the data

The first essential indicator is the nominal resolution d_{min} of the data. This latter is defined as 'the minimum distance at which two features in the corresponding electron density map can be resolved' [5]. It is also defined as d_{min} as the resolution of the X-ray diffraction pattern where, following the Bragg equation: $d_{min}=\lambda\sin\theta_{max}/2$, θ_{max} being the maximum value of θ for reflections contained within the measured data set. Atomic resolution parallels with a d_{min} of about 0.8 Å, for copper radiation.
Generally, a smaller d_{min} shows that smaller structural details are noticeable in the maps. For many crystallographic procedures, such as direct phasing, the resolution of the data set may be a restrictive parameter; Giacovazzo indicates that $d_{min}\approx 1.4$ Å is the resolution threshold below which a small molecule structure cannot be deciphered ab initio[6]; consequently, its precise measurement is important.
In some cases, an overestimation of the nominal resolution can also be identified based on the appearance of the Wilson [7] plot of the data set. In particular, a clear deviation of the plot from linearity at the nominal resolution side is often a good indication

3-2 The completeness of data

An additional factor, the second one, is the completeness of the data set. In reality, the established measure d_{min} may change after the addition or removal of a few reflections. This measure does not reveal the incompleteness of the data set.
Weiss [5] proposed portraying the `effective resolution' based on the nominal resolution d_{min} and the cube root of the completeness of the data set C; $\mathbf{d_{eff}= d_{min}\ C\wedge(-1/3)}$, with $\mathbf{C=(N_{ind} / N_{refl})}$ .This relation is attributed to the fact that missing reflections result in a deterioration of the final model parameters much in the same way as reduced resolution does [8] .
It is possible to guess the number of reciprocal lattice points within the range θ to θ_{max} for a given crystal, the number of reflections N_{refl} is given by, $\mathbf{N_{refl}=33{,}510Vc\ \sin^3\theta_{max}/\lambda^3}$, the number of reflections present N_{ind} may not be congruent with the expected number N_{refl} from the nominal value of d_{min}.
If the number of reflection essentially measured is considerably lower than the expected value, it can be anticipated that the resolution of the analysis will be correspondingly lower although it is not easy to estimate by how much. For example, a nominal d_{min}=1.4 A° data set which is only 70% complete will effectually be a 1.6 A° data set.

To collect additional complete diffraction data sets, a positive side effect of a requirement for authors to report the efficient resolution of their data set instead of the nominal resolution might likewise be an increased motivation. Indicators of accumulated statistical properties are needed since the number of reflections in a crystallographic experiment is high. For a long period of time, the 'merging' R value has been used to measure the consistent regularity of measurements.

3-3 The internal consistency

The merging statistics offer a valuable source of information about the quality of a data set prior to merging equivalent reflections.The most regularly reported descriptor of data quality is the usual merging R factor R_{merge},$\boldsymbol{R_{merge}} = \sum_h \sum_l |\boldsymbol{I_{hl}} - \langle \boldsymbol{I_h} \rangle| / \sum_h \sum_l \langle \boldsymbol{I_h} \rangle$ which measures the spread of n independent measurements of the intensity of a reflection, I_i(hkl), around their average, Î(hkl).These R factors cannot be paralleled with the R factors in refinement (R_{work} or R_{free}),the R_{merge} is fundamentally dependent on the redundancy of the data [5], Low redundancy will always generate a lower R_{merge}, but at the same time lead to less exact data. Diederichs&Karplus [3], show that to give a result that is independent of the average multiplicity N, each term of the numerator has to be adjusted by a factor of $[N/(N-1)]^{1/2}$.The resulting quantity is called R_{meas} (or $R_{r.i.m}$; [5]). These merging R factor (R_{merge} or better $R_{r.i.m}$), are not suitable measures for setting a resolution cutoff [4]. $\boldsymbol{R_{r.i.m}} = \sum_h \sum_l |I_{hl} - \langle I_h \rangle| \sqrt{\left(\frac{N}{N-1}\right)} / \sum_h \sum_l \langle I_h \rangle$ Subsequently, it was indicated that by introducing an additional factor of $(1/N)^{1/2}$into each term of the R_{meas} numerator, the quality of the merged data could also be expected [5], as this explains the expected increase inaccuracy related with averaging N measurements. The resulting quantity is called $R_{p.i.m}$. (The precision-indicating merging R; [5]) which defines the precision of the averaged measurement. $\boldsymbol{R_{p.i.m}} = \sum_h \sum_l |\boldsymbol{I_{hl}} - \langle \boldsymbol{I_h} \rangle| \sqrt{\left(\frac{1}{N-1}\right)} / \sum_h \sum_l \langle \boldsymbol{I_h} \rangle$ .The R factor should deliver the most information when it comes to guessing the performance of a given data set in structure determination, given the fact that the course of structure determination and refinement, averaged intensities or amplitudes are typically used.

The conventional criteria for examining the data resolution limit are the R_{merge} and $R_{r.i.m}$ and $R_{p.i.m}$. Yet, they are not adopted universally and are not required by journals that publish X-ray structures. Lately, (Karplus & Diederichs, [9]) we have advocated that the Pearson correlation coefficient of two 'half' data sets $CC_{1/2}$might be better appropriate than merging R factors for assessing and judging data quality.

Some of the advantages of a correlation constant are that it has a clear and accurate range: +1.0 for a good correlation and 0 for no correlation. $CC_{1/2}$ is commonly close to 1 at low resolution and falls sharply to near zero at higher resolution as the intensities become weaker [10]

4.Analysis of data quality

The results for the practical examples tested are summarized in table 1

Table1:*Details of the two data sets used in the practice.*

Crystal data		Crystal data	
Chemical formula	$C_{20}H_{21}N_5O$	Chemical formula	$C_{20}H_{21}N_5O$
M_r	347.42	M_r	347.42
Crystal system, space group	Monoclinic, *P2/n*	Crystal system, space group	Orthorhombic, $P2_12_12$
Temperature (K)	293	Temperature (K)	293
a, *b*, *c* (Å)	12.2410 (4), 5.9386 (2), 25.2112 (9)	*a*, *b*, *c* (Å)	9.7570 (3), 32.0691 (13), 5.8643 (2)
β (°)	90.317 (3)		
V (Å^3)	1832.69 (11)	*V* (Å^3)	1834.93 (11)
Z	4	*Z*	4
Radiation type	Cu *Kα*	Radiation type	Mo *Kα*

Data 1 : ot-Cu:Slassi et al.IUCrData (2019) **Data 2** : ot-Mo:Slassi et al.IUCrData (2017)

IOP Conf. Series: Materials Science and Engineering **783** (2020) 012005 doi:10.1088/1757-899X/783/1/012005

An interesting feature was observed by comparing the tow data sets, According to a conventional Signal-to-noise ratio, **I/σ(I)>2** criterion:

- The nominal resolution d_{min} of the tow data sets would be 0.84 and 0.7 Å respectively.

- The number of measured, independent and observed [I> 2σ(I)] reflections would be

	N_{meas}	N_{ind}	N_{obs}
Data 1	***5548***	***3487***	***3037***
Data 2	***25352***	***3590***	***2978***

- The completeness increased with higher resolution **Figure1**, It has long been realized that any missing reflection, be it one that has not been measured leads to a deterioration of the structure parameters

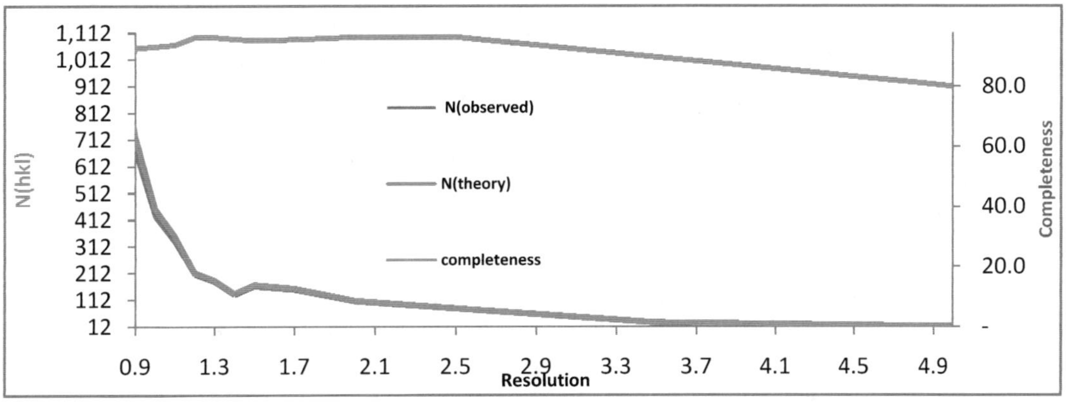

Figure1: *Plots from Excel of completeness ,N_{obs}(hkl)and N_{theo}(hkl)against resolution in Data1*

- The Signal-to-noise ratio I/σ(I) increased with lower resolution (**Figure2** and **Figure3**),I/σ(I) is a good criterion for resolution cutoff.

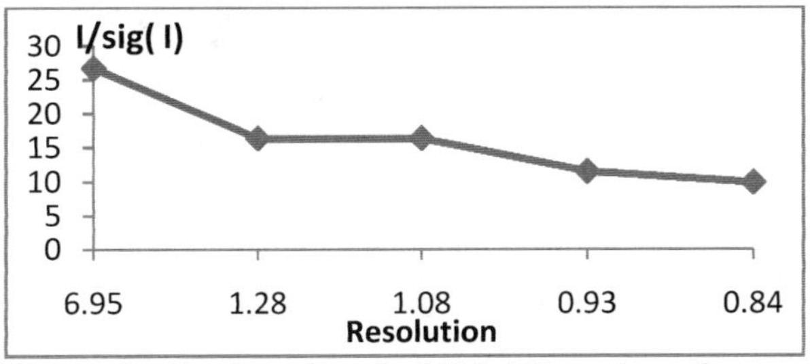

Figure 2: *plot from Excel of I/sig(I) against d_{min} in Data1*

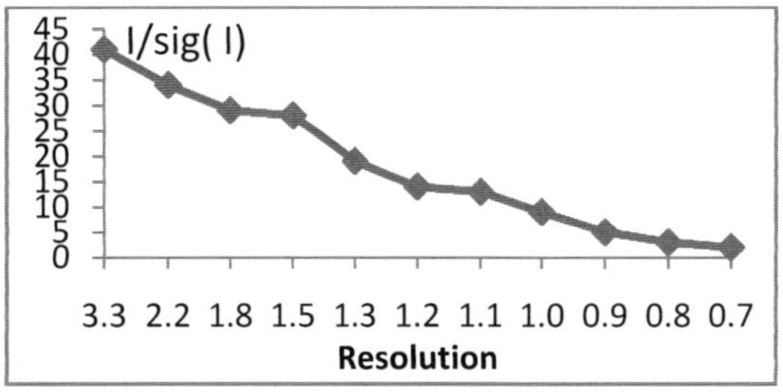

Figure 3: *plot from Excel of I/sig (I)against d_{min} in Data2*

- The R_{merge} or R_{sym} value is generally increased with higher resolution and completeness, overall R_{sym} values of <5%(**Figure 4** for d_{min}>0.9),5-10%(**Figure 5** for d_{min}>0.9) are taken to indicate good, usable quality data [3]. R_{sym} is commonly used to guide decisions during data reduction, such as determining to what resolution data are reliable.

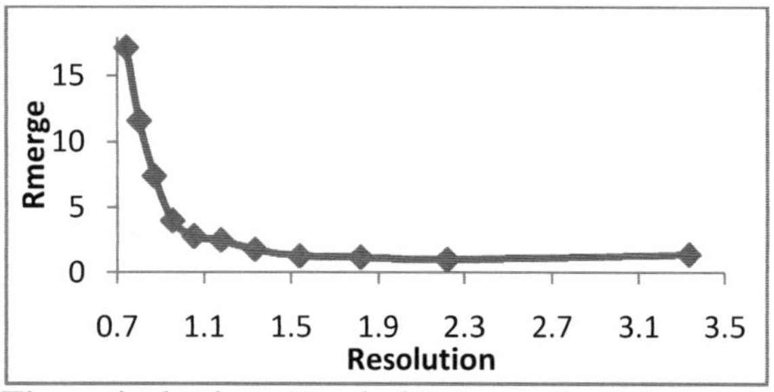

Figure4:*plot from Excel of R_{merge} against d_{min} in Data2*

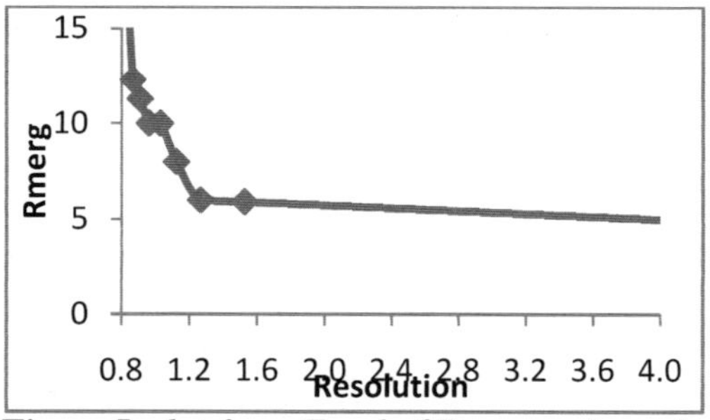

Figure5:*plot from Excel of R_{merge} against d_{min} in Data1*

5.Application of resolution cutoff

How can we decide where to apply a resolution cutoff? To examine these questions, a number of tests were carried out using two data.By comparing refinement against measured data with refinement against simulated data, we can judge the resolution point at which the measured data become no better than simulated data and compare this with the data-processing statistics.
The crystal structure of Data2 was then refined to 0.7A resolution against the observed data, R values (and wR) for the experimental data increase with increasing resolution using all data;

Table2: *Comparison with simulated Data2*

	N_{ind}	N_{obs}	R	wR	S	obs/ para
Data2	3590	2978	0.037	0.10	1.04	12.5
Cutoff	1895	1632	0.035	0.08	1.04	7

A better structure has a lower R values and better data allow obtaining a better structure as shown in **Table2.**Similar results were obtained with example Data1, as shown in **Table3;**

Table3: *Tests of resolution cutoffs using Data1*

	N_{ind}	N_{obs}	R	wR	S	obs/para
Data1	3487	3037	0.084	0.245	1.08	12.8
Cutoff	500	454	0.05	0.15	1.08	2

Choice of high-resolution cutoff is better because it gives lower R-values and better structure

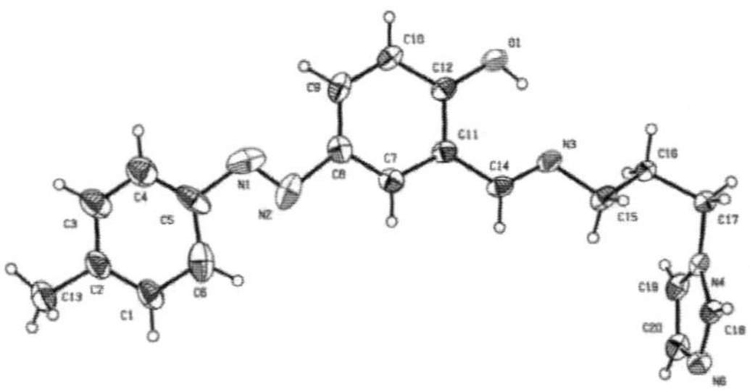

Figure6:The structure of crystal data1

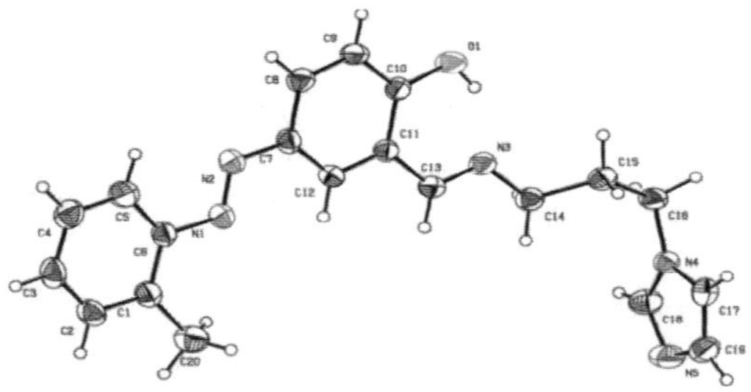

Figure7:The structure of crystal data2

IOP Conf. Series: Materials Science and Engineering **783** (2020) 012005 doi:10.1088/1757-899X/783/1/012005

6. Conclusions

The results can be summarized in the following statements:

- ✓ Merging many data sets usually improves the completeness and signal-to-noise ratio in the data; I/σ(I) is a good criterion for resolution cutoff.
- ✓ Choice of high-resolution cutoff is better because it gives lower R-values
- ✓ Structure quality depends on data quality.

7. Keywords

Diffraction data, crystal structure, resolution, global indicators, R values, cutoff.

8. References

[1] Fox, G. C. & Holmes, K. C.1966. *Acta Cryst.***20**, 886–891.
[2] Kabsch, W.2010. *Acta Cryst.*D **66**, 133–144.
[3] Diederichs, K. &Karplus, P. A.1997. *Nature Struct. Biol.* **4**, 269–275.
[4] Diederichs, K. &Karplus, P. A. 2015 .*Current Opinion in Structural Biology*. **34,** 60–68
[5] Weiss, M. S.2001 . *Appl. Cryst.* **34**, 130–135
[6] Caliandro, R., Carrozzini, B., Cascarano, G. L., De Caro, L.,Giacovazzo, C. & Siliqi, D. 2005a. *Acta Cryst.*D **61**, 1080–87.
[7] Wilson, A. J. C.1949 . *Acta Cryst.* **2**, 318–321.
[8] Evans 2011. *ActaCryst.* D **67**, 282–292
[9] Karplus, P. A. &Diederichs, K.2012. *Science*. **336**, 1030–33.
[10] Evans &Murshudov 2013. *ActaCryst. D* ***69***, 1204–14.

IOP Conf. Series: Materials Science and Engineering **783** (2020) 012006 doi:10.1088/1757-899X/783/1/012006

Inverse geochemical modelling of pit lake waters in the Zeida abandoned mine (Morocco)

L El Alaoui, A Dekayir, M Rouai
ER Geotech, Faculty of Sciences, Moulay Ismail University, Meknes, Morocco

Abstract. In the high Moulouya watershed, the exploitation of the Zeida mine between 1972 and 1985 generated huge masses (millions of tons) of mining waste (Pb, Zn). This activity created large pits currently filled with stagnant water which are in permanent contact with remaining ore deposit.
The interaction of rainwater with rocks and minerals lead to the enrichment of Water Lake with heavy metals.
An inverse geochemical modelling carried out on water sampled in ZA and ZL1 lakes made it possible to highlight the chemical reactions occurring between rainwater and ore deposit and which would control the chemistry of these waters.
KEYWORDS : Contamination, Heavy metals, Inverse Modelling, Zeida, Morocco.in constructing both.

1. Introduction

In water geochemistry, many methods have been used to study, evaluate and characterize the variation of water composition. These methods include multivariate statistical techniques and inverse geochemical modelling [1;9].
PhreeqC inverse geochemical modelling [7] is based on a geochemical model of molar balance, which quantifies the mole transfer of phases (mole of minerals and liquid to be reacted in a solution) taking into account differences in the initial and final composition of the water along a flow path.
At least two chemical analysis of water sampled at different point of the pathway and a set of phases are necessary to perform the calculation].
This study aims to determine the chemical mechanism that controls the chemical composition of ZA and ZL1 pit lakes waters from the Zeida abandoned mine located in high Moulouya.

2. Material and methods

In the eastern part of Morocco, the Zeida abandoned mine is located in a semi-arid to arid climate (300mm/year) with a scarcity of water resources. The pit lakes are used by people to support their domestic and agricultural needs. In this study two samples were collected from ZA and ZL1 pit lakes.
Some of these samples were filtered at 0.22µm using a Nalgene filtration tool and then frozen in a dry ice. The parameters pH, T and C.E of waters were measured during the sampling.

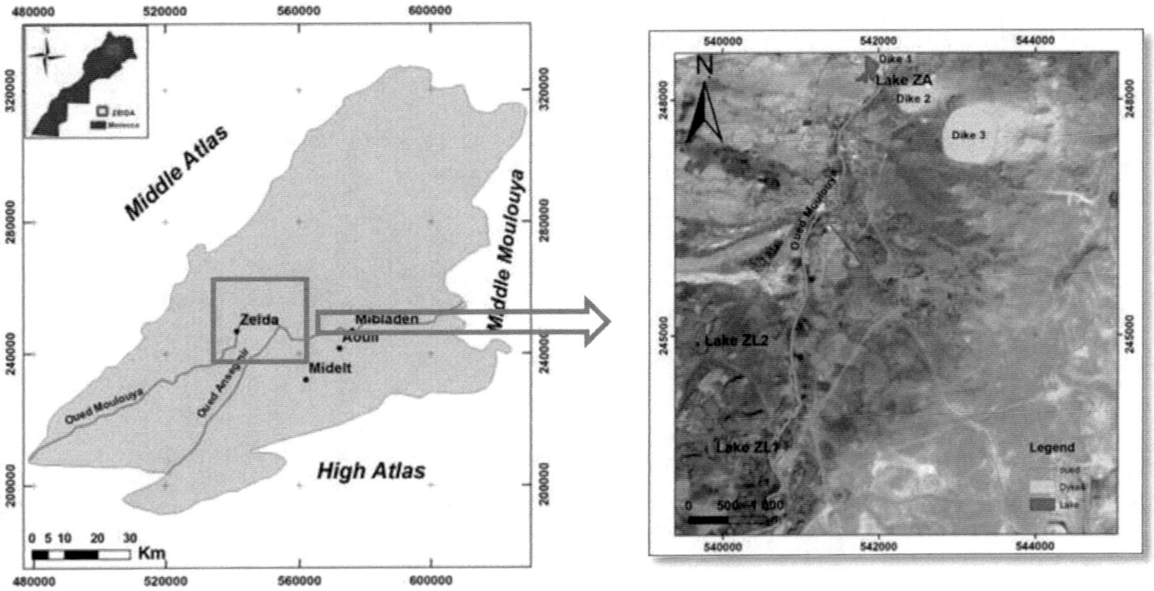

Figure 1. Situation Of the studied lakes in Zeida mining area

For the inverse modelling calculation, PhreeqC has been used [7]. The calculation of the mass balance during water infiltration is the number of moles of minerals that must dissolve or precipitate to account for the difference in composition between the initial water (rain water tab.1) and the final water (lake ZA or ZL1 waters, Tab.2) [5-8].

The analysis of the metallic traces elements were carried out on the filtered water by the ICP-AES technical then the main cations, anions, and the total organic carbon were carried out by ion chromatography (Dionew ICS-1000), and a carbon analyser organic (Shimadzu TOC-VCSH) according to a high temperature catalytic oxidation process [2] in hydrosciences laboratory, Montpellier in France [5]

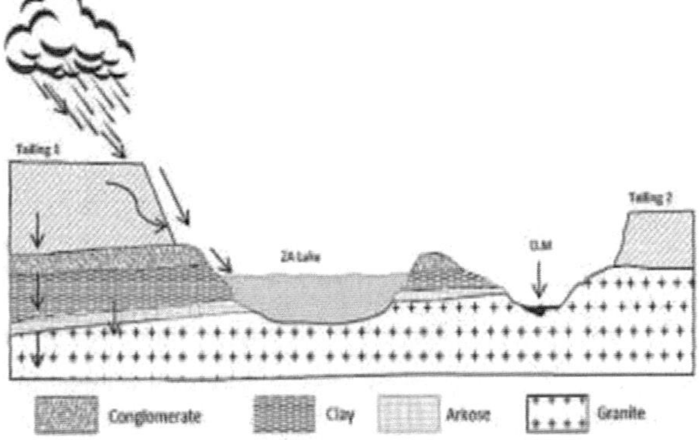

Figure 2. Geological section showing the disposition of tailing (tailing I and II) in relation to pit lakes ZA and Oued Moulouya (O.M)

3. Results and discussion

The chemical data of the studied waters are presented in table 2.These waters are generally alkaline with pH between 9.02 and 9.08. The conductivity (EC) varies between 5630 µS/cm in ZL1 Lake at 16°C as evidenced by the precipitation of salt on the edges of ZL1 Lake. The ZA water shows a weak (pe) compared to that of ZA one reflecting a more reducing environment (Tab.1).
The ZA water show lower values of sodium, potassium, magnesium, calcium, chloride and sulphate than ZL1 water. In ZL1, bicarbonate show high value (1864 ppm) followed by chloride with a value of 2703.79 ppm and the (SO_4^{2-}) with a value of 1397.59 ppm (Tab.2).
The rainwater used in this calculation as an input solution in the system shows an alkaline pH with an average chemical composition given in Table 1. This water is enriched in SO_4, in chloride and calcium with a (pe) of 4 reflecting oxidative conditions.

Table 1. Physical and chemical properties of rainwater (ppm) (Solution 1)

pH	Pe	Ca	Cl	K	NO3-	SO4
7	4	21	9	0.8	1.6	9.5

Table 2. Physical and chemical properties of ZA and ZL1 water (Solution 2) (ppm)

	pH	**Pe**	**Ca**	**Cl**	**K**	**NO_3**	**SO_4^{2-}**	**Al**	**As**
ZA Lake	**9.02**	**0.35**	**29**	**943.12**	**33.12**	**0**	**96.51**	**0.01109**	**0.04389**
ZL1 Lake	**9.08**	**1.189**	**16020**	**2703.79**	**118.37**	**0**	**1397.59**	**0.0024**	**0.09705**

	B	**Ba**	**U**	**Br**	**CH_4**	**Cd**	**Cs**	**Cu**	**F**	**Fe**
ZA Lake	**0.05595**	**0.06437**	**0.02363**	**0**	**82.82**	**-**	**0.000219**	**0.001844**	**0**	**0.003621**
ZL1 Lake	**2.461**	**0.0768**	**0.2764**	**0**	**480.15**	**0.000323**	**0.00333**	**0.000515**	**0**	**0.007661**

	Li	**Mg**	**Mn**	**Pb**	**Rb**	**Sr**	**Zn**	**HCO_3**
ZA Lake	**1.851**	**42.27**	**0.0252**	**0.000022**	**0.02578**	**2.823**	**0.02961**	**450.7**
ZL1 Lake	**11.16**	**187.07**	**0.008253**	**0.000021**	**0.2101**	**3.473**	**0.2899**	**1864**

Hydrochemistry of studied waters: The Piper diagram show the composition of major hydrochemical facies identified for each lake waters ZA and ZL1. From the piper plot (Fig.3), major hydrochemical facies identified were Na^+,K^+,SO_4,Cl^+,NO_3 with minor amount of Ca and Mg.

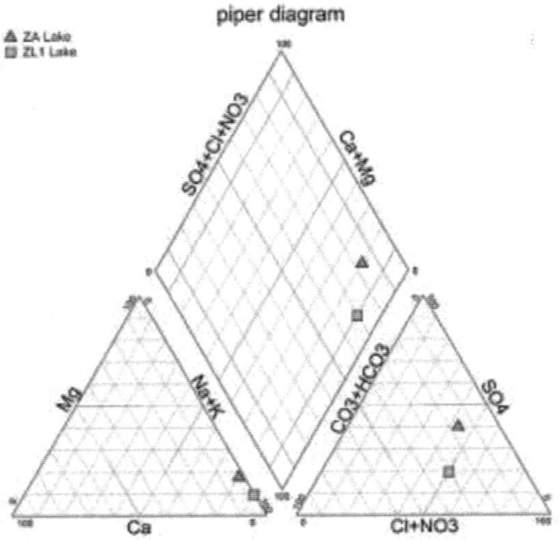

Figure 3. Representation of ZA and ZL1 water compositions in the piper diagram

Water saturation index: Calculation of the saturation index (SI) of waters consists in predicting the mineral phases which must precipitate or dissolve. The PhreeqC aqueous speciation model [7] was used for this. It's equal to log10 of the product of the ionic activity of a mineral component divided by the equilibrium constant for the considered mineral. Indeed, an SI>0 indicates a precipitation in water (saturated) SI<0 indicates an under saturation of the water while an SI=0, indicates a state of equilibrium of the mineral constituents with the aqueous solution.

The values of the saturation index calculated in the two water are shown in (Fig.3).These waters show over-saturation for hematite, aragonite and dolomite and under-saturation of the halite, cerussite, calcite and arsenic phases (Fig.4)

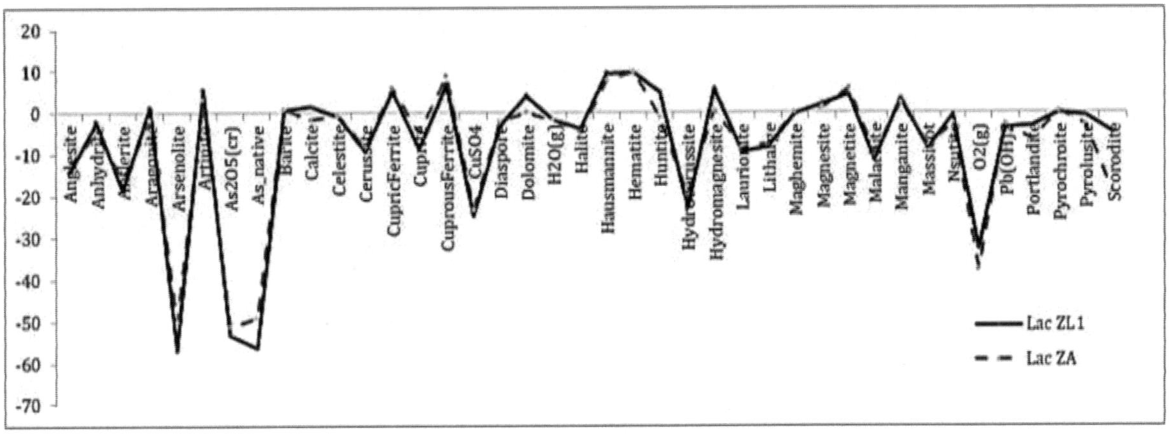

Figure 4. Water saturation index of ZA and ZL1waters

Table 3. Result of the inverse geochemical modelling of ZA water

Mineral/Phases	Calcite	Cerussite	Halite	Hematite	Scorodite	Aragonite
Model-I	-9.86e-05	9.89e-05	2.66e-02	-2.61e-07	5.88e-07	-
Model-II	-9.89e-05	9.89e-05	2.66e-02	-	-	-9.86e-05
Model-III	-	9.89e-05	2.66e-02	-2.61e-07	5.88e-07	-
Model-IV	-9.89e-05	9.89e-05	2.66e-02	-	6.50e-08	-
Model-V	-5.24e+05	5.24e+05	-2.61e-07	-5.88e-07	-	-
Model-VI	-2.67e-02	-1.74e-10	6.54e-08	-	-	
Model-VII	-	9.89e-05	2.66e-02	-6.50e-08	-	-9.89e-05

Table.4. Results of the inverse geochemical modelling of ZA water

Mineral	Anhydrite	Calcite	Cerussite	Halite	Hematite	Scorodite	Aragonite
Model-I	1.08e-01	-1.08e-01	1.08e-01	7.6e-02	-5.8e-07	1.30e-06	
Model-II	1.08e-01	*	1.08e-01	7.6e-02	*	*	-1.08e-01
Model-III	2.23e-04	*	3.21e-04	7.6e-02	-5.8e-07	1.30e-06	*
Model-IV	1.08e-01		1.08e-01	7.6e-02	*	1.38e-07	-1.08e-01
Model-V	2.23e-04	*	3.21e-04	*	-5.8e-07	1.30e-06	
Model-VI	2.23e-04	*	*	7.6e-02	-1.7e-10	1.38e-07	
Model-VII	1.08e-01		1.08e-01	7.6e-02	-5.8e-07		-1.08e-01

Inverse geochemical modelling: The minerals used in the calculation of the inverse geochemical modelling are those present in the study area. The positive values of mass transfer indicate dissolution of the minerals or gases while the negative values indicate a precipitation or degassing.
According to the inverse calculation, the primary mineral phases used are calcite, cerussite, halite, hematite, scorodite, and aragonite.
In ZA water from I to VII models, minerals such as hematite, calcite and aragonite show precipitation, while cerussite, halite and scorodite show dissolution (Tab .3).
The inverse geochemical modelling calculated for ZL1 water, shows 7 (Table 4), from which minerals such as calcite, hematite and aragonite show a precipitation while halite, anhydrite, scorodite and cerussite show a dissolution.(Tab.4)
The calculation of saturation index along the flow path between the input solution (rainwater) and the final solution (water lake) fit perfectly with the inverse calculation data [4].

Conclusion
The saturation index calculations and inverse geochemical modelling applied along the precipitation (rainwater) – pit lakes water pathways show that the chemical compositions of these waters results from the interaction between the rainwater and the primary minerals of rocks and sediments crossed by these waters.
This chemical mechanism is characterized by dissolution of minerals such as anhydrite, cerussite, scorodite, halite and the precipitation of calcite, hematite and aragonite. Metal minerals based on lead (galena) and arsenic show essentially dissolution in these waters.

ACKNOWLEDGEMENT
This work has been supported by the grant in the framework of CNRST Projet Prioritaires N° 34/2015.

References

[1] Belkhiri L, Boudoukha A, Mouni L, and Baouz T 2010 *Geoderma.* **159**(3-4), 390-398.
[2] Casiot C, Egal M, Elbaz-Poulichet F, Bruneel O, Bancon-Montigny C, Cordier M A, and Aliaume C 2009 *Applied geochemistry.* **24** (5), 787-799.
[3] Charlton S R, Macklin C L, and Parkhurst D L 1997 *Water-Resources Investigations Report.* **9**, 7- 4222.
[4] Dhiman S D, and Keshari A K 2006 *Environmental geology.* **50**(8), 1211-1219.
[5] EL Alaoui, L., Dekayir, A. (2018) E3S Web of Conferences **37**,01007.
[6] Kenoyer G J, & Bowser C J 1992 *Water Resources Research.* **28**(2), 591-600.
[7] Parkhurst D L 1999 *Water-Resources Investigations Report* 99-4259. US Geological Survey.
[8] Plummer L N, Parkhurst D L, & Thorstenson D C 1983 *Geochimica et cosmochimica Acta.* **47**(4), 665-685.
[9] Suma C S, Srinivasamoorthy K, Saravanan K, Faizalkhan A, Prakash R, & Gopinath S 2015 Aquatic Procedia. **4**, 986-992.

IOP Conf. Series: Materials Science and Engineering **783** (2020) 012007 doi:10.1088/1757-899X/783/1/012007

Thermophysical characterization of a new clay-based construction material from the Atlas region

A. LKOUEN[1], A. MESKINI[1] , M. LAMRANI[2], A. KHABBAZI[3]

[1]LMIET, ENSEM, Hassan II University - Casablanca, Morocco

[2]LASTIMI, EST, Sale, Mohamed V University- Rabat, Morocco

[3]MEAT-EMDD, EST, Sale, Mohamed V University- Rabat, Morocco

Corresponding author: a.lkouen@gmail.com

Abstract. As part of the valorization of local materials and the search for energy efficiency in the building sector, a thermophysical study was carried out on a new clay-based composite material from the Atlas region reinforced by peanut shells, which constitute an abundant ecological and untapped waste. In this study, a series of samples (100x100x26 mm^3) of this composite are elaborated at different mass fractions of peanut shells. The thermal parameters are estimated by metrological methods such as the steady-state hot plane method and the flash method. The results showed, in addition to the lightness offered by peanut shells to these composites, that the thermal conductivity and thermal diffusivity vary respectively from values 0.623 $W.m^{-1}.K^{-1}$ and 4.8 $m^2.s^{-1}$ for raw clay to the values 0.323 $W.m^{-1}.K^{-1}$ and 3.5 $m^2.s^{-1}$ for the mass fraction of peanut shells $\varpi = 9\%$, which give this material the quality of the insulating walls.

1. Introduction

The building industry remains the largest consumer of energy and among the factors that can reduce this consumption is thermal insulation, therefore our contribution is to develop new building materials with improved thermal performance while giving value to local materials.

A lot of research has been carried out on local materials such as cement stabilized clay studied by H. Ezbakhe et al [1]. El Bakkouri et al [2] presented a thermomechanical study of lightweight concrete with cork or with olive pomace. Khabbazi et al [3] carried out an experimental study of the thermal and mechanical properties of a new insulating material derived from cork and cement mortar. Elhamdouni et al [4] have demonstrated the effect of alfa fibres on the thermophysical characterization of a clay extracted from northern Morocco. Mounir et al [5] have worked on the thermal inertia and thermal properties of clay-plastic composites. The same authors [6] carried out an experimental study of the thermal properties of clay and sheep's wool samples, the results showed that the addition of sheep's wool gives the clay a 45% lightness factors. Cherki et al [7] studied the thermal behavior of an ecological composite material based on gypsum plaster reinforced with cork grains, they found that this composite is thermally better than plaster without additives. Lamrani et al [8] highlighted the effect of olive pomace on an extracted clay from southern Morocco. In another work, the same authors studied the thermal performance of a new consolidated plaster-based material with peanut shells, their results showed that the thermal conductivity is reduced from 0.301 $W.m^{-1}.K^{-1}$ for plaster without additives to 0.141 $W.m^{-1}.K^{-1}$ for plaster with a mass fraction of 20% of the peanut shells, thus proving the possibility to use this new material as a false ceiling [9] .

In this work we are interested in developing and characterizing a new material based on clay and peanut shells which is a renewable natural product and eco-friendly. We have highlighted the influence of the mass proportions of additives on the density and thermophysical properties of the materials produced.

Among the metrological methods used to identify thermophysical parameters, we chose the steady-state hot-plane method developed by Jannot et al [10] to estimate the thermal conductivity of the material and the Flash method to estimate thermal diffusivity [11][12]. Many authors have already used these methods to characterize the thermophysical properties of heterogeneous composite materials, so these methods have shown each time their robustness for insulating materials.

2. Sample preparation

The materials used are:

Raw clay: it is a local natural material coming from the region of the atlas, it can constitute the binding phase of the composite

Peanut shells: it is an abundant, ecological and untapped waste. The dispersed particles of the peanut shells used have sizes between 2,5 mm and 5 mm.

Figure 1. Used materials: a) Clay (b) Peanut shell

The SEM image of a particle of peanut shells (figure 2) shows a microporous morphology that justifies a possibility of air containment improving the thermal insulation of the material.

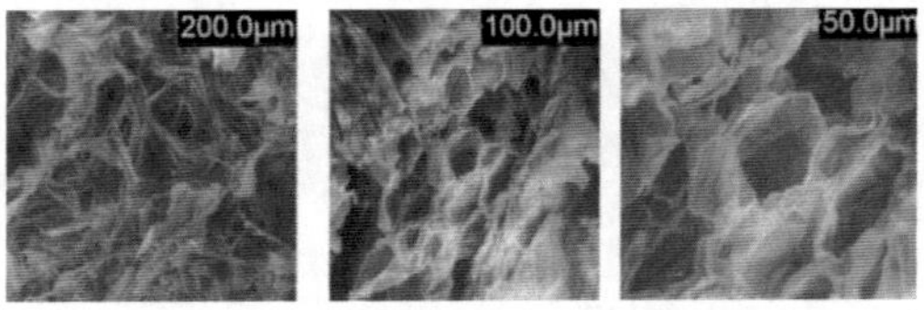

Figure 2. SEM image of a peanut shell particle (CNRST-UATRS)

The samples prepared are composed of a clay matrix containing peanut shells. We found, after a few tests, that the mixing rate that allows the clay paste to be prepared with a normal consistency is

$$q_w / q_c = 0.3 \quad (1)$$

with q_w and q_c are respectively the quantity of water and the quantity of clay.

Samples are prepared by incorporating the peanut shells into the clay paste and then poured into molds. After 24 hours they are removed from the molds and left in the open air for 4 days. To remove moisture from the pores of each of these samples, they are dried in a vacuum draying oven at a temperature of 60 degrees. After drying they are weighed and stored in plastic bags.

The macroscopic aspect of samples with different mass contents of peanut shells is illustrated in Figure 3.

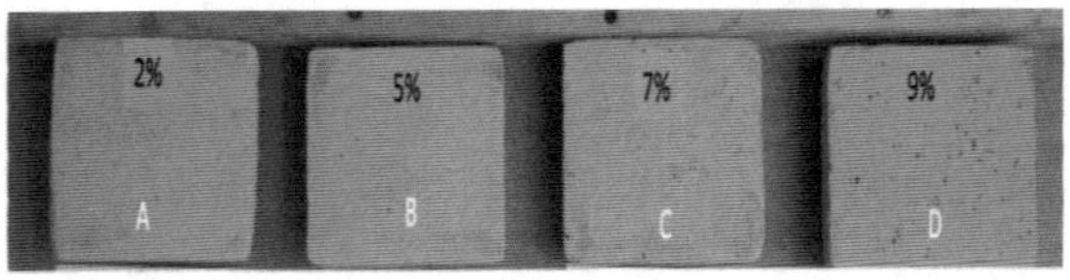

Figure 3. Macroscopic appearance of samples prepared with the proportions 2%,5%,7% and 9% of peanut shells

3. Theoretical approach and experimental description of measurement of thermal properties

3.1. Hot plate method in steady state:

This method makes it possible to characterize the thermal conductivity of the samples[10] , its description is given in figure 4, where a heating element is sandwiched between the sample and an insulating foam, in order to maximize the flow of heat through the sample, the thermocouples are positioned to measure the temperatures T_1 and T_0 at the centers of the upper and lower faces of the sample and the temperature T_2 at the center of the heating element , the role of the two aluminium blocks, which has a fairly high thermal conductivity, is to reach steady state after a reasonable time.

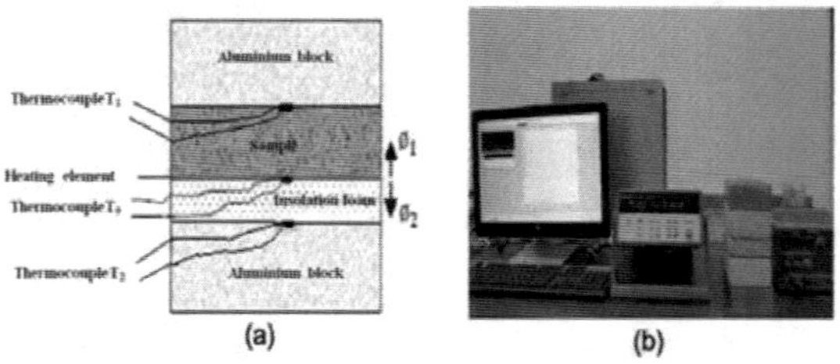

Figure 4. Asymmetric hot plate method in steady state: (a) Principle of the method

(b) Experimental device

We can write:

$$\phi = \frac{U^2}{RS} = \phi_1 + \phi_2 \tag{2}$$

$$\phi_1 = \frac{\lambda}{e}(T_0 - T_1) \tag{3}$$

$$\phi_2 = \frac{\lambda_i}{e_i}(T_0 - T_2) \tag{4}$$

with:

ϕ is the total flow emitted by the heating element by joule effect, $\lambda_i = 0{,}04\, W.m^{-1}.K^{-1}$ and $e_i = 10\, mm$ are successively the thermal conductivity and the thickness of the insulating foam. e is the thickness of the sample, R and S are successively the electrical resistance and the surface of the heating element traversed by an electric current I under the effect of a voltage U imposed on its terminals.

The combination of these equations leads to the expression of the thermal conductivity of our sample:

$$\lambda = \frac{e}{(T_0 - T_1)}\left[\frac{U^2}{RS} - \frac{\lambda_i}{e_i}(T_0 - T_2)\right] \tag{5}$$

3.2. Theoretical models of equivalent thermal conductivity of a two-phase composite material

Theoretical models have been developed in the literature for predicting the equivalent thermal conductivity of a composite material consisting of two phases: continuous phase and dispersed phase [13][14][15]. The expressions of the equivalent thermal conductivity for each model are given by the following relationships:

-Series model:

$$\lambda_{series} = \frac{1}{\frac{1-y}{\lambda_{cont}}+\frac{y}{\lambda_{disp}}} \quad (6)$$

-Parallel model:

$$\lambda_{parallel} = (1-y)\lambda_{cont} + y\lambda_{disp} \quad (7)$$

-Beck model:

$$\lambda_{Beck} = \sqrt{\lambda_{series}.\lambda_{parallel}} \quad (8)$$

- Woodside model:

$$\lambda_{woodside} = \lambda_{disp}^{y}.\lambda_{cont}^{(1-y)} \quad (9)$$

-Maxwell model:

$$\lambda_{Maxwell} = \frac{(1-y)\lambda_{cont}(2\lambda_{cont}+\lambda_{disp})+3y\lambda_{cont}\lambda_{disp}}{(1-y)(2\lambda_{cont}+\lambda_{disp})+3y\lambda_{con}} \quad (10)$$

with: y and λ_{disp} are respectively the volume content and the thermal conductivity of the dispersed phase and λ_{cont} is the thermal conductivity of the continuous phase

3.3. Flash method

This method is used to estimate the thermal diffusivity of solids [11][12], its principle is described in Figure 5:

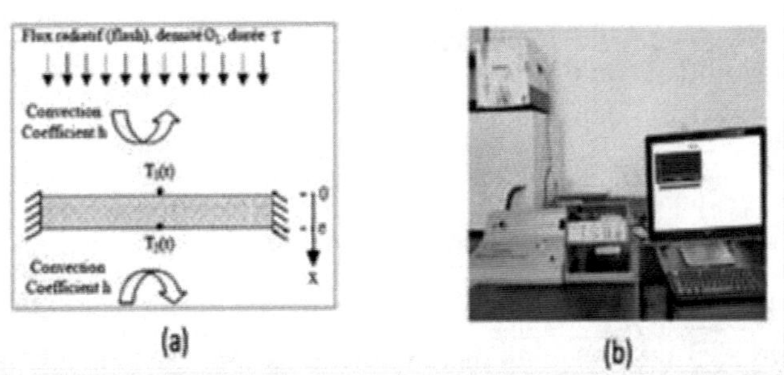

Figure 5. Experimental device of the flash method

(a) Principle of the method (b) picture of the device

A luminous pulse is sent by high power lamps (flash lamps) to the upper surface of the sample. The heat losses on both sides of the sample to its environment are represented by the same convective exchange coefficient h. The lateral sides of the sample are insulated and the luminous flux absorbed by the sample is assumed to be uniform. The temperature evolution over time, at a point on the underside, is monitored by means of a thermocouple or an infrared photodiode.

The modeling of the flash method is based on the resolution of the heat equation in Laplace space. In the sample of thickness e, the equation of heat propagation in Laplace space, in the case of unidirectional heat transfer (1D), is written:

$$\frac{d^2}{dx^2}\theta(x,p) = \frac{p}{a}\theta(x,p) \quad (11)$$

$\theta(x,p)$ is the Laplace transform of the temperature rise $T(x,t) - T_0$, with T_0 is the ambient temperature.

The Laplace transform of the temperature elevation on the underside of the sample is determined using quadruples formalism [16] by the expression:

$$\theta_2(p) = \frac{\phi_0}{p} \frac{(1 - e^{-\tau p})}{\lambda k sinh(ke) + 2h cosh(ke) + h^2 \frac{\sinh(ke)}{\lambda k}} \quad (12)$$

With:

ϕ_0 **:** The thermal flux arriving on the upper surface of the sample (W. m $^{-2}$)

τ : the short time of the flash (s)

p : Laplace's parameter

a, λ are respectly the thermal diffusivity and the thermal conductivity of the sample and $k = \sqrt{\frac{p}{a}}$.

The numerical inversion of the expression is ensured by the algorithm of De Hoog[17] thus giving the theoretical expression of temperature as a function of time $T_{theo}(t)$.

The Levenberg–Maraquart algorithm [18][15] is used to estimate parameters reducing the quadratic error between the experimental temperature recorded at the center of the sample and theoretical expression given by the complete model

$$\Psi = \sum_{i=0}^{N} (T_{exp}(t_i) - T_{theo}(t_i))^2 \quad (13)$$

these two algorithms are programmed in MATLAB

4. Results and discussion

In order to take account of the measurement error, three measurement tests are carried out for each sample and the arithmetic mean of these three tests is adopted.

4.1. Density of the samples

Knowing the dimensions and dry mass of the samples, we determined the average density of each sample (Table 1) with a relative error less than 0.3%.

The apparent density of the composite diminishes with increasing fraction of peanut shell which makes it possible to offer lightness to the material produced. On the other hand, the use of the law of mixtures for a two-component medium allows the volume fraction *y* of the peanut shells in each sample to be deduced:

$$\rho_{sample} = y \rho_{peanut\ shells} + (1 - y) \rho_{clay} \quad (14)$$

Table 1. Apparent Density of the samples produced and Volume fraction of the additive in each one

Sample	Apparent density ρ (Kg/m³)	Volume fraction of peanut shell *y*

D	1455,22	0,335
C	1662,83	0.213
B	1746,80	0.164
A	1905,31	0,072

4.2. Thermal conductivity

Using the steady-state asymmetric hot plane method, three measurement tests are carried out to statistically validate the results.

Figure 6 and Table 2 show the variation in the average thermal conductivity of the samples as a function of the shells content.

Table 2. Results of characterization of the average Thermal Conductivity of the prepared samples

Sample	Volume fraction of peanut shell *y*	Thermal conductivity λ ($W.m^{-1}.K^{-1}$)
Raw clay	0	0,623
A	0,072	0,525
B	0,164	0,435
C	0,213	0,373
D	0,335	0,323

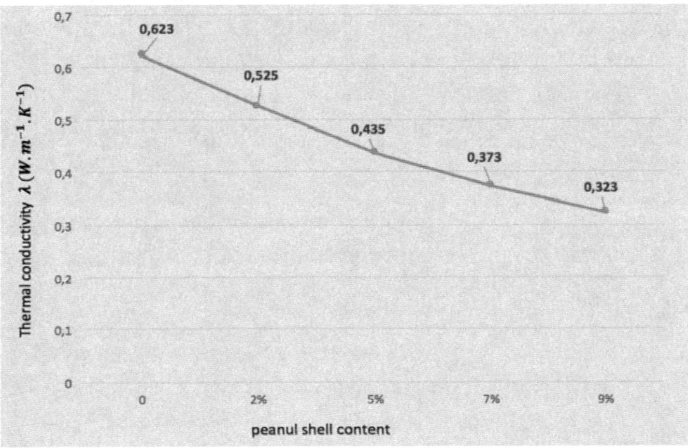

Figure 6. Evolution of the thermal conductivity of the material produced as a function of the percentage of peanut shells

From these results, it can be noted that:

- The measurement error does not exceed 3%, which shows the robustness of the method used

- The average thermal conductivity of the samples decreases with increasing shell content, making the processed material more and more thermally insulating.

4.3. Comparison of conductivity with theoretical models of equivalent thermal conductivity

The figure 7 shows the plot of the curves obtained by applying the theoretical models: series, parallel, Beck, Maxwell and Woodside as well as the experimental curve of thermal conductivity obtained by the hot-steady-state method.

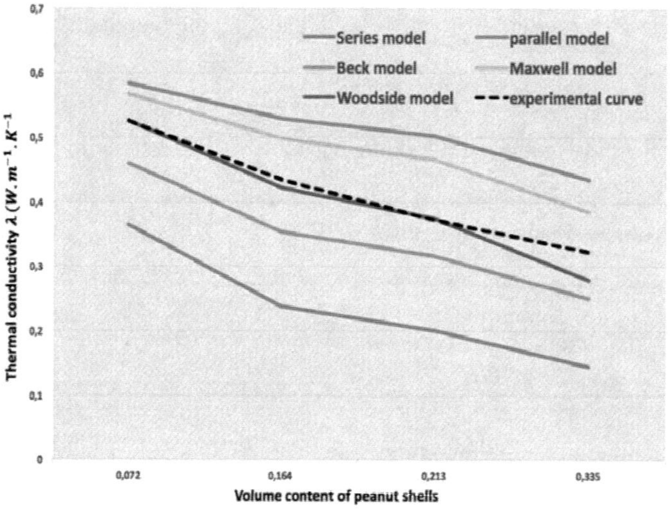

Figure 7. Thermal conductivity of the composite developed as a function of the volume content of peanut shells compared to theoretical models

When observing the curves in the figure, we can already see that the experimental results lie between the lower limit (series model) and the upper limit (parallel model) of thermal conductivity. It is clear that the prepared samples are not represented by the maxwell model because the peanut shells are not spherical, whereas the geometric woodside model is the closest to representing these samples, reflecting the geometric mean of a random distribution of additives in the clay matrix.

4.4. Practical interest of elaborate composites

On the basis of the experimental results obtained, the following two elements can be introduced to value each composite in comparison with the raw clay:

-Lightness factor of the composite compared to the raw clay:

$$F_{\text{Lightness}} = 100 \times (1 - \frac{\rho_{composite}}{\rho_{clay}}) \qquad (15)$$

-Energy saving that can be obtained using two walls of the same thickness and subjected to the same temperature gradient, one consisting of the composite and the other consisting of pure clay:

$$E_{Saving} = 100 \times (1 - \frac{\lambda_{composite}}{\lambda_{Clay}}) \tag{16}$$

The preceding expressions are applied to evaluate the results obtained of the factor of lightness and energy saving for the different elaborated samples compared to clay only (figure 8).

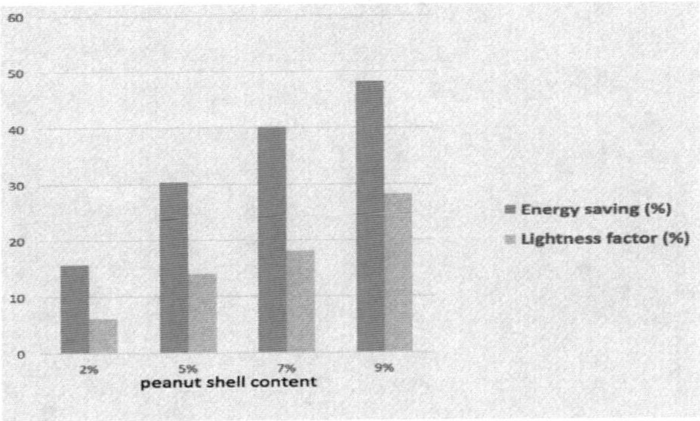

Figure 8. Lightness factor and energy saving of the different materials compared to the raw clay for each mass percentage of peanut shells

We notice that when we increase the percentages of peanut shells, these two parameters increase and the energy gain approaches 50% for a mass fraction of 9%.

4.5. Thermal diffusivity

The flash method allows to identify the parameters: Thermal diffusivity a, thermal flux ϕ_0 and the heat exchange convection coefficient h , by minimizing the quadratic error between the experimental curve and the theoretical curve (Figure 9) given by the complete model and programmed under MATLAB. The results of thermal diffusivities obtained are summarized in Table 3, the measurement uncertainty of the thermal diffusivity is about 3%. The residue, defined as the difference between the two curves, is almost flattened as shown in Figure 9, which confirms that until the date t = 1400 s the 1D model is still valid. Figure 10 illustrates the reduced sensitivity curves for the three parameters to be estimated. Analysis of these curves shows that thermal diffusivity can be reliably estimated at a time around 300 s, while the other two parameters can only be estimated at longer times.

Table 3. Results of characterization of the average thermal diffusivity of the prepared samples

Sample	Volume fraction of peanut shell y	Thermal diffusivity $a.10^{-7}$ ($m^2.s^{-1}$)
Only clay	0	4,80
A	0,072	4,34
B	0,164	3,90
C	0,213	3,74
D	0,335	3,23

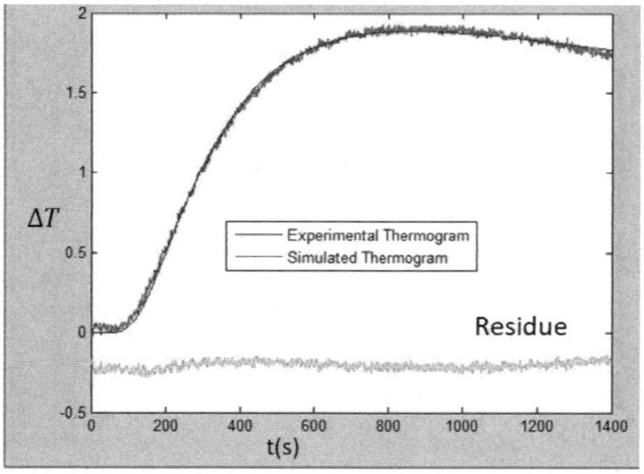

Figure 9. Experimental and theoretical thermogram after minimization, corresponding to the sample D

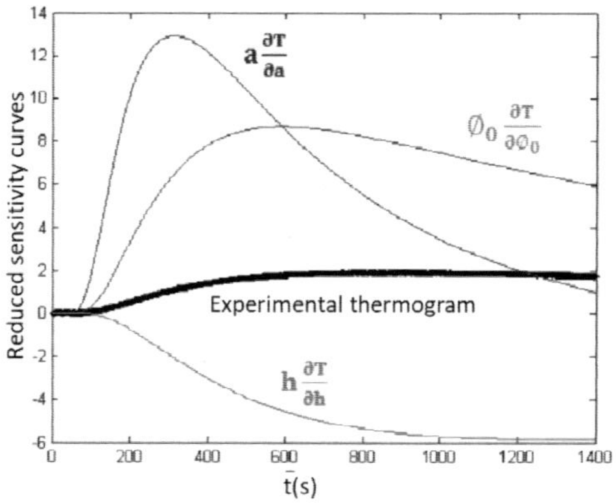

Figure 10. Curves of the reduced sensitivities of the three parameters to be estimated corresponding to the sample D

Conclusion

This study underlined that the thermal insulation quality of clay, as a traditional building material, is significantly improved by the addition of peanut shells, thus offering better energy efficiency. However, this work has focused on the influence of additive content, which opens up a possibility to study the impact of the internal structure of the composite (the shape of the additives and their dispersion in the binder) on its thermal behavior. It will also be interesting to carry out an in-depth study on the optimization of the mechanical and acoustic properties involved in the choice of building materials.

References

[1] Ezbakhe H,Bousad S and ElBakkour A (2001)Thermal study of the Cement Stabilized Earth used in Construction in Northern Morocco *Renewable and Sustainable Energy Reviews* **69** 69–72.

[2] ELBakkouri A,Ezbakhe H, Ajzoul T and ELBouardi A (2005) Thermomechanical study of the concrete with cork and concrete with olive waste *Journées Internationales de Thermique* **12** 307–310.

[3] Khabbazi A, Garoum M, E. Supérieure, D. T. De Salé, and D. Génie (2005) Experimental study of thermal and mechanical proprieties of a new insulating material based on cork and cement mortar *Association for the Advancement of Modelling and Simulation Techniques in Enterprises AMSE Periodicals* **74(7)** 73-80.

[4] Elhamdouni Y,Khabbazi A,Benayad1 C,Mounir S and Dadi A (2017) Thermophysical and mechanical characterization of clay bricks reinforced by alfa or straw fibers *IOP Conf. Series: Materials Science and Engineering* 186 4–11.

[5] Mounir S,Khabbazi A,Maaloufa Y, and Khaldoun A (2016) Thermal Inertia and Thermal Properties of the Composite Material Clay-plastic *Research Journal of Applied Sciences, Engineering and Technology* **12(5)** 507–515.

[6] Mounir S,Khabbazi A,Khaldoun A, Maaloufa Y (2015) Thermal Inertia and Thermal properties of the composite material Clay-wool Thermal inertia and thermal properties of the composite material clay – wool *Sustainable Cities and Society* **19** 191–199.

[7] Cherki A,Remy B,Khabbazi A,Jannot Y, and Baillis D (2014) Experimental thermal properties characterization of insulating Experimental thermal properties characterization of insulating cork – gypsum composite *Constr. Build. Mater* **54** 202–209.

[8] Lamrani M,Khalfaoui M,Laaroussi N and Khabbazi A (2018) Thermal characterization of a new effective building material based on clay and olive waste *MATEC Web of Conferences* **149 (02053)** 1–6.

[9] Lamrani M, Laaroussi N, Khabbazi A,Khalfaoui M,Garoum M and Feiz A (2017) ecological building material based on peanut shells and plaster Experimental study of thermal properties of a new ecological building material based on peanut shells and plaster *Case Studies in Construction Materials* **120** 1-14

[10] Jannot Y, Felix V, and Degiovanni A (2010) A centered hot plate method for measurement of thermal properties of thin insulating *Measurement Science and Technology* **21** 035106 (8).

[11] Parker WJ,Jenkins RJ,Butler CP, and Abbott GL (1961) Flash method of determining thermal diffusivity, heat capacity, and thermal conductivity *Journal of Applied Physics* **32 (9)** 1679–1684.

[12] Degiovanni A, Laurent M,Degiovanni A, and Laurent M (1986) A new thermal diffusivity identification technique for the flash method *Revue Ahys Appl* **21** 229-237

[13] Maxwell JC (1873) A treatise of Electricity and Magnetism,Clarendon Press,Oxford,**1** part II Chapter IX

[14] Woodside W and Messmer JH (1961)Thermal Conductivity of Porous Media. I. Unconsolidated Sands *Journal of Applied Physics* **32** 1-13.

[15] Bal H, Jannot Y, Gaye S, and Demeurie F (2013) Measurement and modelisation of the thermal conductivity of a wet composite porous medium Laterite based bricks with millet waste additive *Constr. Build. Mater* **41** 586–593.

[16] Degiovanni A, Batsale JC, and Maillet D (1996) Measurement of the longitudinal diffusivity of anisotropic materials - Overview of the techniques developed at LEMTA *Revue générale de thermique* **35** 141–147.

[17] De Hoog FR, Knight JH and Stokes AN (1982) An improved mathod for numerical inversion *Society for Industrial and Applied Mathematics* **3(3)** 357–366 .

[18] Marquardt DW (1963) An Algorithm for least-squares estimation of nonlinear parameters *J.Soc. Iindust Appl Math* **11(2)** 431-441.

IOP Conf. Series: Materials Science and Engineering **783** (2020) 012007 doi:10.1088/1757-899X/783/1/012007

Acknowledgments

Special thanks are extended by LMIET, ENSEM of Casablanca, to the research teams of MEAT-EMDD, EST Sale for their help in carrying out this study.

Sorption of contaminants into thermoplastic pipes: *A Critical Review*

L. ELAMINE[1], A. ELMALIKI[2], F.MAJID[3]

1: Team of Research Applied to Polymers, ENSEM, Morocco, elamine.lamyae@gmail.com
2: Team of Research Applied to Polymers, ENSEM, Morocco, a.elmaliki@ensem.ac.ma
3: Team of Research Applied to Polymers, ENSEM, Morocco, majidfatima9@gmail.com

Abstract. For several decades, polymeric materials are increasingly used in various fields, particularly in the distribution of drinking water thanks to the various advantages represented by polymers such as light weight, low cost, easier to install and without any problem of corrosion compared to metal pipes. One of the main concerns of researchers is the sorption of the constituents of drinking water through the pipe. Many studies have found that the mass transfer of a liquid into the polymer has consequences such as the change of crystallinity, the yellowing of the polymer, decrease of mechanical resistance and the swelling of the polymer. The aim of this study is to carry out a review about the sorption of the constituents of liquid transported into the material in order to summarize what is already done in this field. We started with the theoretical study of sorption then the experimental one by showing the different methods used to measure the diffusivity of a contaminant towards the polymer then we enumerated the empirical and analytical models developed for predicting diffusivity and solubility coefficients and finally we showed the influence of this transfer on the mechanical characteristics and the morphology of thermoplastic pipes.

1. Introduction

For several decades, polymer materials have seen their use increase in various fields, particularly in the distribution of drinking water in pipes or fittings. Polyethylene and PVC are the most commonly used materials for the transport of drinking water in Morocco thanks to the various advantages they offer such as lightness, low cost, ease of installation and the absence of corrosion problems compared to metal pipes.

If the soil and water environment are contaminated then the buried water pipes will come into contact with the pollutants and despite the fact that plastics are said to be impermeable to chemicals but the molecules of some products can penetrate. Therefore, understanding the sorption of contaminants in the pipe is now a requirement to assess the future use of the distribution system.

The sorption process begins with an adsorption of the molecules that settle on the inner surface of the pipe and then an absorption of these molecules into the pipe which is managed by the partition coefficient K and finally diffusion of sorbed contaminant molecules through the amorphous area within pipe materials.

Several factors influence this mass transfer, including environmental conditions, pipe properties and contaminant properties. Contaminants diffuse into the amorphous regions of the material, the cross-linking, the bulk density of the pipe, the size shape and polarity of the contaminants can also influence polymer interactions[1].

In this study, a review of available literature was developed about the sorption of contaminants into thermoplastic pipes in order to serve as a source of background information and to summarize what is already done in this topic area. The specific objectives of this study were: 1-intializing the theoretical study of sorption 2- listing the different methods used to measure the diffusivity of a contaminant towards the polymer 3- identifying the empirical and analytical models developed for predicting diffusivity and solubility coefficients and finally we showed the influence of this transfer on the mechanical characteristics and the morphology of thermoplastic pipes.

2. Theory

The background between the pipe and the drinking water is a heterogeneous background so diffusion is controlled by fick's law:

$$\frac{\partial C}{\partial t} = D\frac{\partial^2 C}{\partial x^2} \qquad (1)$$

The solution of this equation for the case of diffusion in polymers is given by Crank and Park [2] using the following assumptions:

The diffusion coefficient D is constant (does not depend on the concentration)
Initial conditions : t=0 ; C(x)=f(x)=0
Boundary conditions: t>0; C(x=0)=C1 = C∞
C(x=L)=C2= C∞

$$\frac{C(t,x)}{C\infty} = 1 - \frac{4}{\pi}\sum_{n=0}^{\infty}\frac{1}{(2n+1)}\exp\left(\frac{-D(2n+1)^2\pi^2}{l^2}t\right)\sin\left(\frac{(2n+1)\pi}{l}x\right) \qquad (2)$$

We have $Mt = \int_0^l C(t,x)dx$ (3)

So the simplified formula for diffusion in polymers is:

$$\frac{Mt}{M\infty} = 1 - \sum_{n=0}^{\infty}\frac{8}{\pi^2(2n+1)^2}\exp\left(\frac{-D(2n+1)^2\pi^2 t}{l^2}\right) \qquad (4)$$

To determine the diffusion of a contaminant in a thermoplastic tube, a parameter is calculated which is the diffusivity and is estimated by calculating the slope of the curve α drawn from the equation (5)

$$\frac{Mt}{M\infty} = f(\sqrt{t}) \qquad (5)$$

$$D = \pi\left(\frac{l\,\alpha}{4}\right)^2 \qquad (6)$$

Solubility is also a parameter that allows us to define the ability of a solute to dissolve in a polymer and is calculated according to the following formula:

$$S = \frac{M\infty - M0}{\mathrm{Mt{=}0}}\rho\mathrm{polymère} \qquad (7)$$

With M∞ is the mass of the contaminant in the saturated polymer, M0 is the initial mass of the contaminant in the polymer which is equal to 0 and Mt=0 is the mass of the polymer, ρ is the bulk density.

3. Methods for measuring diffusivity

In the literature, several experimental techniques are used to study transport phenomena or more precisely the estimation of penetrant diffusivity in polymers like nuclear magnetic resonance (NMR), laser interferometry, gravimetric method and Fourier transform infrared spectroscopy (FTIR).

Britton et al. [3]studied the diffusion and permeability of pollutants through HDPE using the gravimetric method. Scherillo et al.[4] studied water diffusion in PEEK and PEI using the gravimetric method and FTIR by comparing the two. Moylan et al.[5] used a quartz crystal microbalance to study the solubility of water vapour in four different polyimides at 100% relative humidity and 22°C and Kumar et al.[6] studied the sorption and desorption of aromatic molecules in PU and PMMA.
The nuclear magnetic resonance technique has also been used to study polymer/solvent transfer such as toluene in polystyrene, water in nylon, water in glass-filled polyesters, polyvinyl acetate glues in wood, 1,4 dioxane in a polybutadiene elastomer and several organic solvents in PMMA.[7][8][9][10]
Laser interferometry was used by Saenger [11] and Durning et al to study the diffusion of methanol and n-methyl-2-pyrrolidone in polymethyl methacrylate (PMMA). [12]
Fourier transform infrared spectroscopy has also been widely studied. Safa et al .[13] have used it extensively in their work to study ester sorption kinetics in polypropylene films or bottles. Riquet et al.[14] used it to study the sorption of olive oil in polypropylene and determine the concentration profiles in the film thickness. Cava et al. [15] used it to determine the diffusion of flavours through an extruded PolyEthylene film. Cotugno et al.[16] also adapted this approach for the study of water/epoxy systems and finally Ichikawa et al.[17] used this spectroscopic technique to study water sorption in different polymer films.
The two methods most used and recommended by scientists are the gravimetric method and Fourier Transform Infrared Spectrometry (FTIR).

3.1. The gravimetric method:

The gravimetric method is one of the most widely used methods for studying the sorption of a liquid in a solid material, which consists in monitoring the evolution of the mass of a sample immersed in this medium. There are two types of gravimetry measurement: static and dynamic methods.

The protocol consists first of all in preparing the samples in advance, then in putting them in contact with the chosen solution. Over time, the sample is removed, wiped, dried and weighed using a microbalance to calculate the difference in mass before and after contact[18].
The collected measurements make it possible to draw a curve equation (5) and thus to find the diffusivity D which represents the slope of the curve.
Figure 1 above shows an example of sorption kinetics of amyl acetate sorption in polypropylene at different concentrations at T=23°C

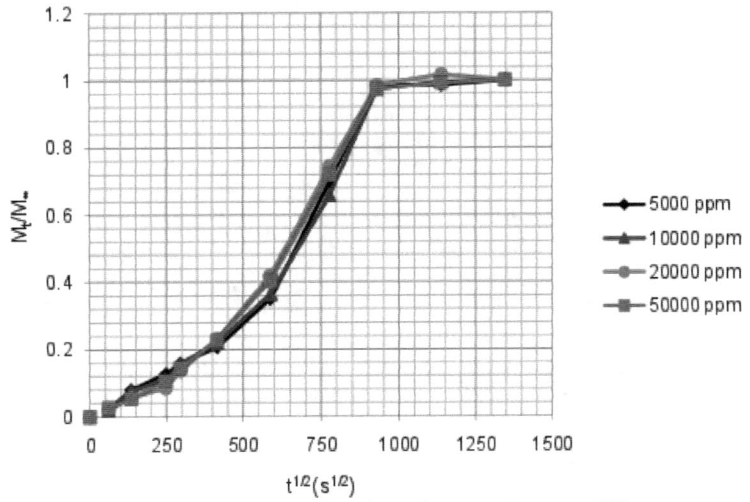

Figure 1 : kinetics of amyl acetate sorption in polypropylene at different concentrations at T=23°C,[18]

3.2. The FTIR method

Fourier Transform Infrared Spectroscopy (FTIR) is based on the absorption of infrared radiation by the analyzed material. It allows, through the detection of the characteristic vibrations of the chemical bonds, to carry out the analysis of the chemical functions present in the material.

Infrared spectrometry is the measurement of the decrease in the intensity of radiation passing through a sample as a function of wavelength. This intensity measurement at each wavelength leads to a spectrum characteristic of the product under study.

The spectrum obtained represents $I/I0=f(\tau)$ with I: absorbed intensity, I0: initial intensity and $\tau=1/\lambda$

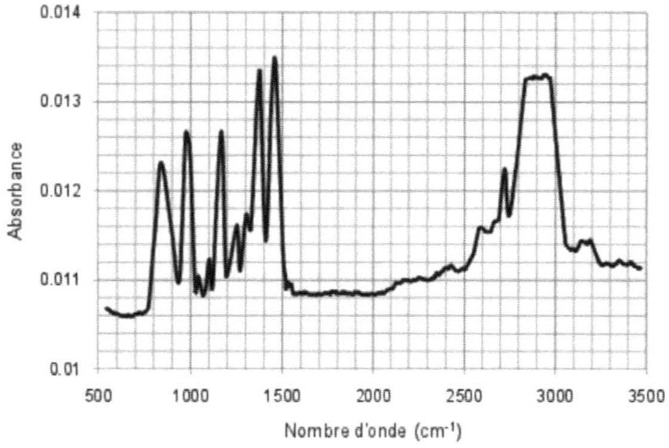

Figure 2 : virgin polypropylene spectrum , [18]

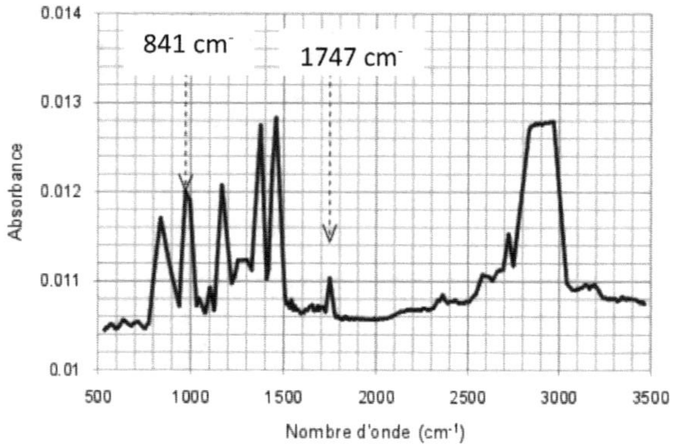

Figure 3 : spectrum of polypropylene in contact with amyl acetate at T=23°C [18]

To quantify the absorbed product, the ratio of the areas Ar of absorbance at time t is plotted on A∞ as a function of the square root of the ageing time:

$$\frac{Ar}{A\infty} = f(\sqrt{t}) \qquad (8)$$

$Ar = \frac{At}{App}$ (9) With At is the peak absorbance area of amyl acetate at time t and App is the area of the characteristic peak absorbance of polypropylene.

A∞ is the ratio of areas to equilibrium

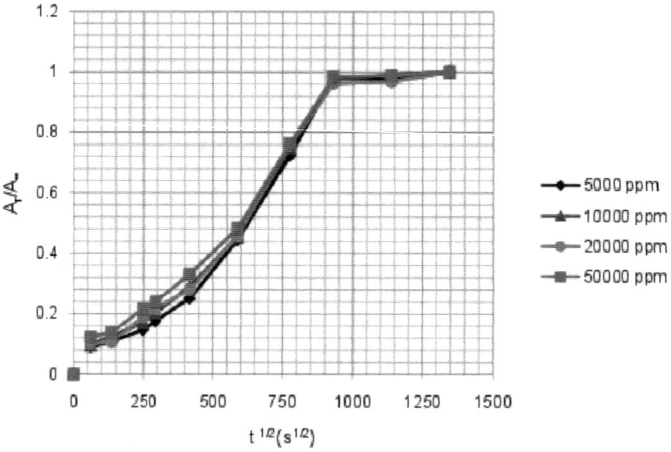

Figure 4 : evolution of area ratios during sorption of amyl acetate in polypropylene , [18]

The diffusion coefficient is obtained by calculating the slope of the linear part of the curve.
FTIR offers the possibility to simultaneously measure the permeability of various penetrants and through the analysis of band shape changes and band shift it is possible to evaluate the different types of interactions that can be established between the polymer and the penetrant at different sorption stages. Thanks to FTIR, sorption can also be easily and accurately evaluated through very thin films [15].

4. Empirical and analytical models:

Several authors have studied the sorption of contaminants through thermoplastic pipes used for the transport of drinking water. Before outlining the models developed by these authors, we will first identify the types of contaminants and the thermoplastic pipes used to do so.

4.1. Contaminants:

The selected contaminants are components commonly found in contaminated hazardous sites and the selection are based on their frequency of detection in contaminated environments [19][20]
Tang et al.[1] assumed that some important properties of contaminants influence polymer interactions like polarity μ, molar volume Mv , molecular weight Mw , water solubility Sw , the density of contaminant ρc and the log form of octanol water partition coefficient logKow..

The table below give the properties of the 14 contaminants selected:

Table 1 :Properties of contaminants, [1]

Contaminant	μ debye	M_v cm^3/mol	M_w g/mol	S_w (%)	P_c g/cm^3	Logk_{ow}
Acetonitrile	3.92	53.5	41.05	100	0.786	-0.34
Benzaldehyde	2.80	120.5	106.12	0.3	1.041	1.48
Benzyl alcohol	1.71	125.1	108.14	4.0	1.041	1.10
1-butanol	1.66	96.2	74.12	7.4	0.809	-0.30
2-butanone	2.76	91.4	72.11	22.3	0.805	0.29
1,2-dichloromethane	1.9	79.01	98.96	0.87	1.25	1.45
Dichloromethane	1.14	60.6	84.93	1.303	1.326	1.25
MTBE	1.36	119.1	88.15	5.1	0.740	1.24
2-propanol	1.56	77.8	60.10	100	0.789	0.05
2-propanone	2.88	73.1	58.08	100	0.789	-0.24
Trichloromethane	1.01	74.4	119.37	0.729	1.492	1.97
Toluene	0.36	117.7	92.14	0.0526	0.867	2.73
m-Xylène	0.30	135.9	106.16	0.0161	0.864	3.20
p-Xylène	0.00	135.8	106.16	0.0162	0.867	3.15

4.2. Types of thermoplastic pipes:

PVC pipe: PVC is an amorphous thermoplastic with a glass transition temperature (Tg) of 82°C. Because its Tg is higher than the ambient temperature, PVC is a glassy and rigid plastic.[21]
At present, there are three types of PVC pipes used in Morocco: unplasticized PVC-U used for drinking water, overchlorinated PVC-C used for the installation of cold and hot water and bioriented PVC.
PE pipe: There are many types of PE pipes with different chemical composition and operating conditions, but these materials have always been classified as PE. All PE pipes have a Tg below zero,

so the material is flexible at room temperature and below ground temperature. All PE pipes are manufactured from PE resins with different densities, crystallinities and Tg values:

Low density:

PELD 0.910 g/cm^3 < ρ < 0.925 g/cm^3 , Tg= -20 °C

Medium density:

PEMD 0.926 g/cm^3 < ρ < 0.940 g/cm^3 , Tg= - 70°C

High density:

PEHD ρ ≥ 0.941 g/cm^3 , Tg= - 140°C

PPR pipe:

Random polypropylene (PP-R) is a widely used material for pipe applications. To protect the polymer against oxidative degradation during service, it is stabilised with phenolic antioxidants. Polypropylene is a semi-crystalline thermoplastic that is semi-rigid and highly resistant to abrasion with a Tg equal to -10°C and a density 0.9 g/cm3.

4.3. Model of Aminabhavi 1999:

Aminabhavi et al. [22]studied the sorption, desorption and permeability of contaminants through HDPE geomembranes using the gravimetric method using the following formula:

$$Ct = \frac{Wt - W0}{W0} \times \frac{1}{M} \tag{10}$$

With Wt is the mass at t, W0 the initial mass of the sample and M the molar mass of the solvent and thanks to these experimental data he developed an expression of these three parameters based on Arrhenius' law which takes into consideration the effect of temperature.

$$X = X_0\, e^{-\frac{Ex}{RT}} \tag{11}$$

With Ex is the activation energy that has been obtained by the least squares procedure.

4.4. Model of chao2007:

Chao et al. [23]determined the diffusion and solubility coefficients of organic solvents through HDPE geomembranes using 2 methods: the ASTM F739 permeation experiment and the immersion experiment, after which he proceeded with numerical analysis to solve the one-dimensional fick equation:

$$Cz(z,t) = kCf(1 - \frac{Z}{L} - \sum_{n=1}^{\infty} \frac{2}{n\pi} \exp\,(\frac{-Dt(n\pi)^2}{l^2}) \sin\,(\frac{n\pi}{l} z)) \tag{12}$$

Based on the following conditions:

C(0)=S= KCf

C(l)=0

t=0, Cz=0

Where K is the solubility coefficient and the Cf is the equilibrium concentration of the permeate.

The initial conditions are the data we had experimentally.

The simulation showed that the immersion method is inappropriate and that ASTM F739 describes better the boundary conditions.

4.5. Model of Tang 2013:

Tang et al.[1] predicted the solubility and diffusion coefficients of contaminants through HDPE. His model was based on the properties of the contaminants and the pipe. The regression method was adopted to select the variables and the linear multiple regression was used to develop the equations. The model was based on experimental data found by Whelton et al.[24].

The validation of the model was done by comparing the results of the model with experimental data from the literature.

The model is as follows:
Diffusivity = 103.10 - 0.840µ - 0.085 Mv + 2.263logKow - 98.066 ρb
Solubility = 0.6734 - 0.00852µ - 0.00111 Mv + 0.0377 logKow - 0.577 ρb

5. Influence of sorption on the pipe:

The pipes used in water distribution networks are subject to environmental constraints, their lifespan depends a lot on the retention of antioxidants by the material. In dry air, polyethylene can retain its properties for several decades. However, when polyethylene pipes are used in networks and chlorine is present in the water, chlorine can diffuse and penetrate, causing the material to lose its antioxidants.

Several authors have studied the influence of the presence of chlorine in water on the pipe. Dear et al.[25] studied the effect of chlorine on polyethylene pipes in water distribution networks. He found that the lifetime of a pipe is determined by the formation of the brittle layer on the inner surface of the pipe and the formation of this layer is promoted and accelerated in the presence of chlorine in the water because of its diffusion in the pipe, antioxidants are consumed. And this changement of physical properties included a reduction of yield stress, elongation to failure, and molecular mass. He has been shown also that because of high chlorine levels the expected 50-years life of HDPE pipes can be reduced and he confirms that if the pipes are located in well-drained soil without containing industrial chemicals the lifespan can be extended more than 50 years.
Whelton et al. [26] have elaborated an accelerated aging with chlorinated water with 45 mg free chlorine for 141 days and found that it had no effect on the surface properties of MDPE-based PEX-A pipes. However, these aging conditions affected the properties of HDPE, HDPE and PEX-B pipes that were aged in the laboratory by the formation of carbonyl bonds on the surface and the significant loss of the oxidative induction time OIT.

6. Conclusion

A review was developped about sorption of contaminants into thermoplastic pipes, we initiate the methods to measure the diffusivity and the models available in the litterature and finally their effect on the pipe but we observe that informations and studies about this phenomenan are few and models that take into account the different factors that influence the diffusion of contaminants through the pipes were missing. The study of the influence of this transfer has highlighted just the degradation of the pipe (the OIT) but data concerning the effect of the transfer on the mechanical characteristics of the pipe, on melting and glass transition temperatures and finally on the morphology of the pipe after contact are also missing.

7. References

[1] J. Tang, D. L. Gallagher, M. Asce, A. M. Dietrich, and A. M. Asce, "Predicting Permeation of Organic Contaminants into Polyethylenes," no. February, pp. 205–212,(2013).
[2]. J. Crank, "THE MATHEMATICS OF DIFFUSION," (1975).
[3]. L. N. Britton, R. B. Ashman, M. Tejraj, and P. E. Cassidy, "Permeation and Diffusion of Environmental Pollutants through Flexible Polymers," vol. **38**, pp. 227–236, (1989).
[4]. G. Scherillo, M. Petretta, M. Galizia, P. La Manna, and P. Musto, "Thermodynamics of water sorption in high performance glassy thermoplastic polymers," vol. **2**, no. May, pp. 1–16, (2014).
[5]. S. Jose, G. T. Division, and H. Junction, "Solubility of Water in Polyimides : Quartz Crystal Microbalance Measurements," pp. 87–92.

[6]. H. Kumar, "A study of sorption / desorption and diffusion of substituted aromatic probe molecules into semi interpenetrating polymer network of polyurethane / polymethyl methacrylate," vol. **46**, pp. 7140–7155, (2005).
[7]. S.Blackband and P.Mansfield "Diffusion in liquid-solid systems by NMR imaging,", pp. 49–52. (1986)
[8]. L. A. Weisenberger, "NMR Imaging of Solvent Diffusion in Polymers," pp. 1117–1126, (1989).
[9]. L. A. Weisenberger and J. L. Koenig, "NMR Imaging of Case I1 Diffusion in Glassy Polymers," vol. **27**, pp. 55–57, (1989).
[10]. W.P.Rothwell and D.R Holecek, "NMR Imaging: Study of Fluid Absorption by Polymer Composites," vol. **22**, no. 1, pp. 241–247, (1984).
[11]. K.L.Saenger and H.M.Tong, "Laser Interferometry : A Measurement Technique for Diffusion Studies in Thin Polymer Films," vol. **37**, no. 6, pp. 432–435.(2004)
[12]. H. M. Tong and K. W. Lee, "A Study ," pp. 4234–4248, (1995).
[13]. B. L. Safa and B. Abbes, "Experimental and Numerical Study of Sorption / diffusion of Esters into Polypropylene Packaging Films," pp. 55–64, (2002).
[14]. A. M. Riquet, N. Wolff, S. Laoubi, J. M. Vergnaud, and A. Feigenbaum, "Food and packaging interactions : Determination of the kinetic parameters of olive oil diffusion in polypropylene using concentration profiles," pp. 37–41 , (2013)
[15]. A. Lo, R. Catala, R. Gavara, D. Cava, and J. M. Lagaro, "On the applicability of FT-IR spectroscopy to test aroma transport properties in polymer films," vol. **23**, pp. 551–557, (2004).
[16]. S. Cotugno, D. Larobina, G. Mensitieri, P. Musto, and G. Ragosta, "A novel spectroscopic approach to investigate transport processes in polymers : the case of water ± epoxy system," vol. **42**, pp. 6431–6438, (2001).
[17]. K. E. N. Ichikawa, T. Mori, H. Kitano, M. Fukuda, A. Mochizuki, and M. Tanaka, "Fourier Transform Infrared Study on the Sorption of Water," pp. 2175–2182, (2001).
[18]. O. ZAKI, "Contribution à l'étude et à la modélisation de l'influence des phénomènes de transferts de masse sur le comportement mécanique de flacons en polypropylène," (2008).
[19]. K. Chao, P. Wang, and B. Wang, "Determining the permeability of organic solvents through PVC pipes using a direct SPME model," Polym. Test., vol. **56**, pp. 110–114,(2016).
[20]. J.C.Joo , J.Y.Kim and A.M.Asce, "Estimation of Mass Transport Parameters of Organic Compounds through High Density Polyethylene," no. May, pp. 790–799, (2005).
[21]. A.J.Whelton and T.Nguyen, "Contaminant Migration From Polymeric Pipes Used in Buried Potable Water Distribution Systems : A Review," pp 679-751 (2013).
[22]. T. M. Aminabhavi and H. G. Naik, "Sorption r desorption , diffusion , permeation and swelling of high density polyethylene geomembrane in the presence of hazardous organic liquids," (1999).
[23]. K. Chao, P. Wang, and Y. Wang, "Diffusion and solubility coefficients determined by permeation and immersion experiments for organic solvents in HDPE geomembrane," vol. **142**, pp. 227–235, (2007).
[24]. A. J. Whelton et al., "Contaminant Diffusion , Solubility , and Material Property Differences between HDPE and PEX Potable Water Pipes," no. February, pp. 227–238,(2010).
[25]. S. K. Campus, "Effect of chlorine on polyethylene pipes in water distribution networks," vol. **220**, pp. 97–111, (2006).
[26]. A. J. Whelton, M. Asce, A. M. Dietrich, A. Asce, D. L. Gallagher, and M. Asce, "Impact of Chlorinated Water Exposure on Contaminant Transport and Surface and Bulk Properties of High-Density Polyethylene and Cross-Linked Polyethylene Potable Water Pipes," no. July, pp. 559–569, (2011).

Acknowledgments
A special thanks to the authors that were responsive to my requests and questions.

IOP Conf. Series: Materials Science and Engineering **783** (2020) 012010 doi:10.1088/1757-899X/783/1/012010

Infrared active modes in double walled CBN nanotube: theoretical study

A. M. Nassir[1], M. Boutahir[1], F. Fergani[1], B. Fakrach[2], AH. Rahmani[1], H. Chadli1, A. Rahmani[1]

[1]Laboratoire d'Etude des Matériaux Avancés et Applications (LEM2A), Université Moulay Ismail,FSM-ESTM, BP 11201, Zitoune, 50000 Meknes, Morocco
[2]Laboratoire de Physique Théorique et Appliquée, Université Sidi Mohammed Ben Abdellah, Faculté des Sciences Dhar El Mahraz Fez, Morocco

***Corresponding author e-mail: mourad.boutahir@gmail.com**

Abstract. This work focuses on a hybrid system consist of single-wall carbon nanotube inside single-wall boron nitride nanotube (CBN nanotube). The intertube interactions between the outer boron–nitride nanotube and inner carbon nanotube are described using a van der Waals interaction. The infrared spectra of CBN nanotubes are calculated in the framework of spectral moments method. The evolution of the infrared spectra of CBN nanotube as a function of their diameters and lengths are discussed. The obtained results are useful for the interpretation of the experimental infrared spectra of CBN nanotube.

1. Introduction

Boron nitride nanotubes (BNNTs) is a similarly structured to that of carbon nanotubes (CNTs) [1, 2] in which carbon atoms are replaced entirely by equal numbers of nitrogen and boron atoms, arranging in a hexagonal lattice. The BNNTs and CNTs share some similar chemical and physical properties [3,4,5], but the electronic properties of CNTs are different from those of BNNTs, when CNTs can be semiconducting or metallic depending on the rolling direction [6] and BNNTs are an excellent insulator [7] with a wide bandgap (5–6 eV) irrespective of the number of walls, diameters and chiralities [6, 8]. In addition, the mechanical properties of CNTs and BNNTs are different because the B-N bond is ionic [1]. These numerous excellent properties of BNNTs and CNTs make them widely studied by experimental and theoretical workers and a candidate for many innovative applications in nanotechnology.

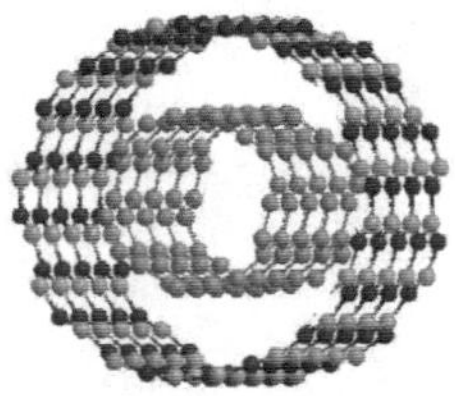

Figure 1. Structure CNT@BNNT.

Similar to carbon nanotube, BNNT can be confined different molecules [9,10]. In addition, under appropriate conditions C60@BNNT provide an effective material for preparing a coaxial structure

composed of an inner CNT and an outer BNNT. Experimentally, hybrid nanostructure CNT@BNNT was observed by high resolution bright-field transmission electron microscopy imaging [11]. This hybrid system can offer an exciting opportunity to form a conducting nanowire inside an insulating BNNT. The infrared spectrum of the synthesized BNNTs and CNTs are dominated by two strong characteristic bonds located around 800 - 820 cm-1 and 1360-1400 cm-1 for BNNTs [12,13,14,15,16] and around 840-900 cm-1 and 1560-1600 cm-1 for CNTs [6,17].
The Raman scattering and infrared absorption are the most powerful tools for investigating the vibrational properties of carbon and BN nanomaterials in relation to their structural and electronic properties [18, 16]. In a previous theoretical work [19], we studied the Raman spectra of C70 and C60 fullerenes inside BNNTs where these molecules adopt a linear arrangement for BNNT diameter lower than 1.45 nm. The calculation of Raman spectra shows that the behaviour of the breathing-like modes (BLM) with the pod diameter was clearly modified in peapod. To obtain a complete assignment of the CBN nanotube, we follow the same methodology but in the case of the infrared spectroscopy in the framework of spectral moments method, where the inner molecule is the carbone nanotube and the outer is BBNT. Our work supports the infrared spectra of (armchair@armchair) CBN nanotubes as a function of their lengths and diameters. The results are useful to understand the experimental infrared data of CBN nanotube.

This paper is organized in the following way. In section 2 we present the used model and method. Results of calculations are the aim of section 3, the final section is conclusion.

2. Computational method

A CBN nanotube consists of the inner single-wall carbon nanotube (SWCNT) trapped inside a single-wall boron nitride nanotube (SWBNNT) nanotube host. Representative structure of optimized (5,5)@(10,10) CBN nanotube is shown in Fig. 1.

A Lennard-Jones potential: $U_{LJ} = 4\varepsilon[(\sigma/r)^{12} - (\sigma/r)^{6}]$ is used to describe the van der Waals interactions between the outer and inner nanotubes, where r is the atom-atom distance, the values of the Lennard-Jones εii parameter is given by 4.116, 6.281 and 2.635 meV, whereas the σ_{ii} parameter is 0.3453, 0.3365 and 0.3369 nm, for boron, nitrogen and carbon atoms, respectively [20–22]. The parameters σ_{ij} and εij between different atoms are calculated by the following Lorentz-Berthelot rules: $\varepsilon_{ij} = (\varepsilon_i\ \varepsilon_j)1/2$ and $\sigma_{ij} = (\sigma_i + \sigma_j)/2$.

The interlayer distance d between the outer and inner nanotube is close to 0.34 nm, this value obtained by minimum energy calculations. Which is in good agreement with that calculated by the density functional theory method [18].

The SWBNNT B-N intratube interactions are described using the same force constant used in recent calculations of the infrared spectra of SWBNNTs [16]. The SWCNT C–C intratube interactions are described by using the same force constants set that the one used in our calculation of the Raman spectrum of isolated SWCNT [23]. The dynamical matrix of SWCNT inside SWBNNT is calculated by block. The system under study contains a large number of atoms, the diagonalization of their dynamical matrix requires long computing time or fails. The most effective way of computing directly the infrared spectra without diagonalizing the dynamical matrix is the spectral moments method [24]. In order to enhance the IR response of both CNT and BNNT nanotubes, dynamical effective charges on tied atoms were fixed at q A=+1 and q B=−1 for a given A-B bond. Please follow these instructions as carefully as possible so all articles within a conference have the same style to the title page. This paragraph follows a section title so it should not be indented.

3. Results and discussions

In all of this study we apply a periodic conditions on the unit cells of nanotubes to obtain the infinite system, also we consider the common Z axis of inner and outer nanotubes to be along the z-axis and the X axis of each nanotubes to be along the x-axis of the laboratory reference frame. The symmetries of the infrared-active modes are directly derived from the polarized IZ(ω) and IX (ω) spectra.
In a first section, we present the results of calculated infrared spectra for infinite hybrid system of CNT@BNNT of armchair@armchair chirality. Next, we report the boron nitride nanotube, carbon nanotube and hybrid system diameter dependence on the frequencies IX(ω) infrared active modes.

Then we also study the influence of the finite length of the inner and outer zigzag tubes on their infrared responses.

3.1. Infinite length of CBN nanotube

We report the calculations of the infrared responses of (10,10)@(15,15)CBN with their corresponding single (10,10)CNT and (15,15)BNNT. The results are displayed in Fig. 2 for three frequency regions: (i) Below 600 cm^{-1} where the breathing-like modes (BLM) dominate, (ii) Intermediate region between 600 and 1000 cm^{-1} where the intermediate-like modes (ILM) are located, and (iii) above 1000 cm^{-1} where the tangential-like modes (TLM) are dominated.

Concerning the $I_Z(\omega)$ polarization, in both BLM and ILM regions no infrared bands are observed in CBN nanotube and also in the corresponding isolated nanotubes. The TLM region is dominated by two bands coming from the corresponding isolated tubes.

Concerning the $I_X(\omega)$ polarization, in the BLM region, the infrared spectrum of the (10,10)@(15,15)CBN displays peaks located at 144 and 237 cm^{-1}. The frequencies of the breathing modes in isolated (10,10) and (15,15) armchir tubes are calculated at 140 and 233 cm^{-1} respectively. Our calculations state a systematic upshift (4 cm^{-1}) of the frequencies of the breathing-like modes with respect to the frequencies of the breathing modes in SWBNNTs and SWCNTs, respectively. In the ILM region, the same behaviour in BLM region is observed. We observe an upshift (2 cm^{-1}) of the infrared bands in CBN nanotube with respect to their positions for single nanotubes. In the TLM region, there are two active modes around 1393 and 1587 cm^{-1}. They are associated with the in-phase and counter-phase coupled motions of the tangential modes of the inner and outer tubes.

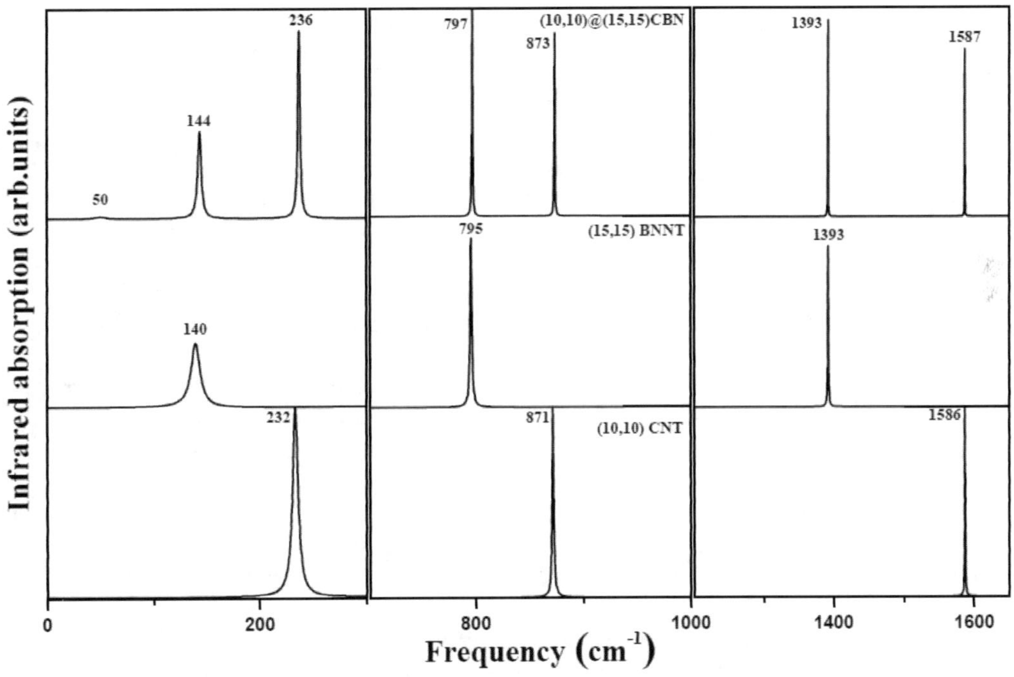

Figure 2. Calculated $I_Z(\omega)$ (dashed lines) and $I_X(\omega)$ (solid lines) of (10,10)@(15,15) CBN nanotube with their corresponding isolated nanotubes.

3.2. Diameter dependence

To illustrate the diameter dependence of the infrared-active modes, the calculated IX(ω) spectra are displayed for three values of the diameter 1.35, 2.03 and 2.71 nm of (10,10), (15,15) and (20,20) SWCNT inside 2.05, 2.74 and 3.42 nm of (15,15), (20,20) and (25,25) SWBNNT, respectively (Fig. 3)

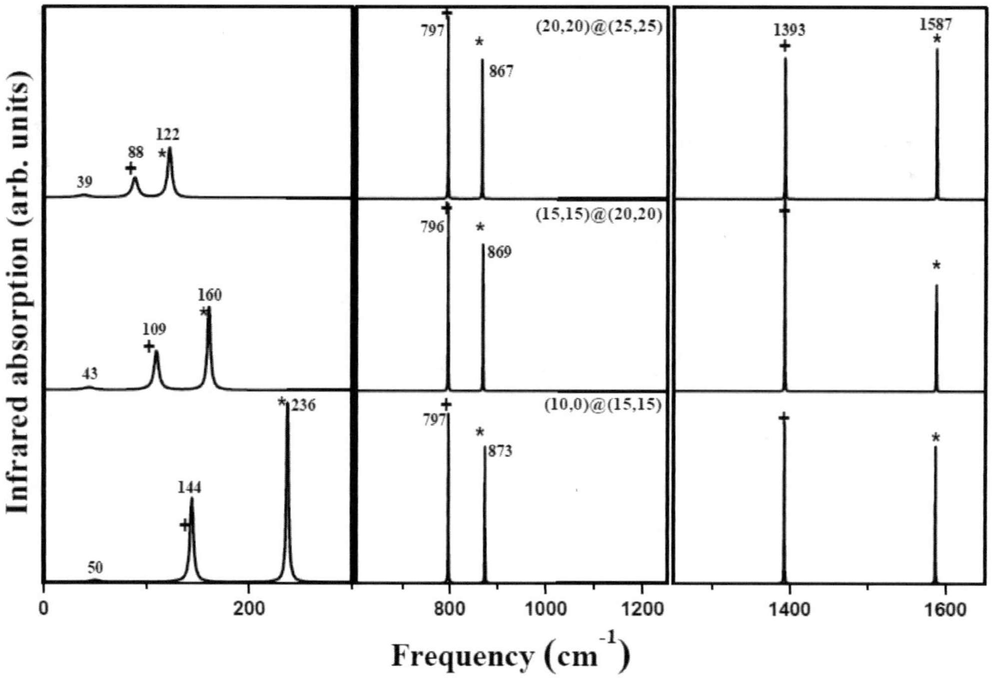

Figure 3. $I_X(\omega)$ calculated IR spectra of infinite (n,n)@(n+5,n+5) CBN nanotube, for n = 10, 15 and 25 from bottom to top in BLM (left), ILM (middle) and TLM (right) regions. The stars and plus give the position of the active modes of carbon and boron nitride isolated tubes, respectively.

In the intermediate mode region, all spectra are dominated by two strong bands around 800 and 870 cm^{-1}. These bands are associated with the counter -phase and in-phase coupled motions of the intermediate modes of the outer and inner tubes.

In the TLM region, spectra are dominated by two strong bands around 1393 and 1587 cm^{-1} when the diameter increased. They are associated with the in-phase and counter-phase coupled motions of the tangential modes of the inner and outer tubes.

In the BLM region, the results indicate two modes low frequency (LF-BLM) and high frequency (HF-BLM). We observe a frequency downshift with increasing tube diameter, for their modes, whereas the intensities decrease. One can see that the BLM modes are strongly dependent on the diameter in comparison with ILM and TLM modes.

We have reported in figure 4 the BLM frequencies for CBN system versus diameter D. From these results, and for CBN in the diameter range 0.6–2.6 nm, phenomenological relations which describe the diameter dependence of the IR-active-mode frequencies have been derived.

The dependence of the frequency of the LF-BLM mode as a function of tube diameter is described by an a/D scaling law with the a parameter depending on the mode symmetry (see Table 1) but the HF-BLM mode is well fitted by the phenomenological relation:

$\omega=a/D+b/D^2+c$

Where D is the outer nanotube diameter. The a, b, and c parameters are given in Table 1 for armchair.

It must be emphasized that this previous expression has a useful application for deriving the diameter of tubes from the experimental infrared-active-mode wave numbers. No physical meaning is attached to the values of the different parameters of the fit.

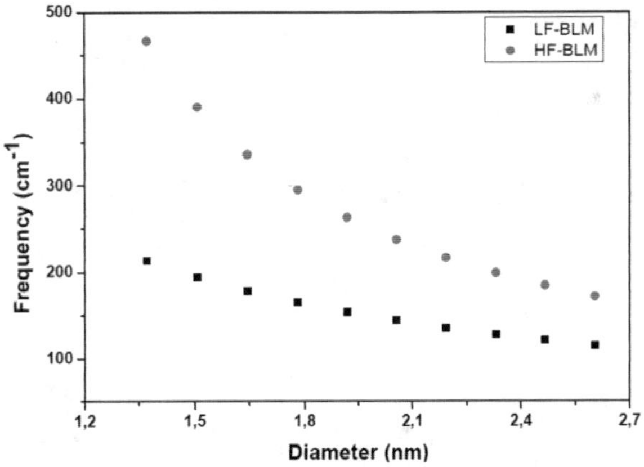

Figure 4. Diameter dependence on the wavenumber I_X infrared active modes for SWBNNT@SWCNT in the low wavenumber region.

Table 1. The a, b, and c parameters for the a/D and a/D + b/D^2 + c dependences of the modes as a function of the tube diameter D.

Modes	Parameters
LF-BLM	a=283 nm cm^{-1}
HF-BLM	a=110 nm cm^{-1} b=-199 nm^2 cm^{-1} c=940 cm^{-1}

Considering the diameter dependence of the specific modes in BLM regions, we present in Fig. 5 the eigendisplacement vectors of BLM lines in a (5,5)@(10,10) CBN nanotube obtained from the direct diagonalization of the dynamical matrix. The eigenvector displacements of the modes located around 212 and 458 cm^{-1} (see figure 4 b and d) are assigned to hybridation between the modes of the outer BN (208 cm^{-1}) and inner carbon (464 cm^{-1}) nanotube (see figure 4a and 4b). All this modes correspond to a mixing of tangential and radial displacement of atoms, as shown in Fig. 5. There are two nodes along the circumference, and all atoms move inward (outward) for the radial component and upgrade (downgrade) for the tangential one on both sides of the two nodes.

3.3. CBN nanotubes with a finite length

Because experiments have been shown a short portions of the inner tube inside the outer tube, we investigate the dependence of the infrared spectra on the lengths of the inner (Li) and outer (Lo) tubes. Thanks to spectral moments method, we can calculate the infrared spectra for finite (9,0)@(18,0) CBN nanotube. In this study, we consider five cases when Li/Lo ratio can take 0.25, 0.5, 1, 1.5 and 2. (figire. 6).

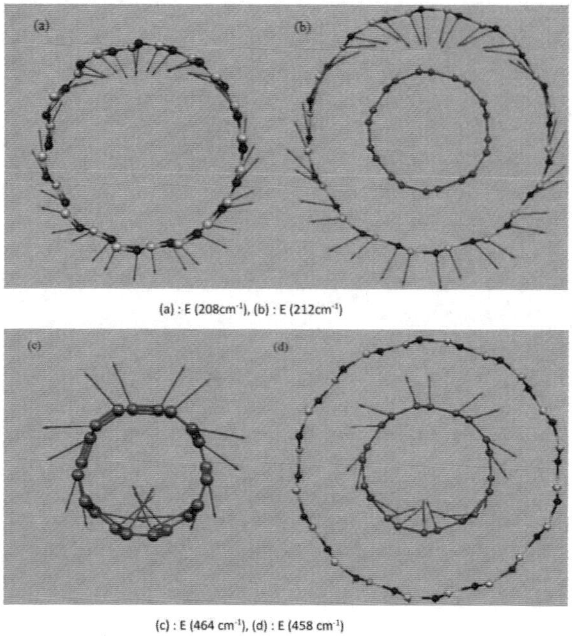

Figure 5. Infrared-active-mode atomic displacements for selected normal modes in the (a) (10,10) BNNT, (c) (5,5) CNT and (b,d) (5,5)@(10,10) CBN nanotube.

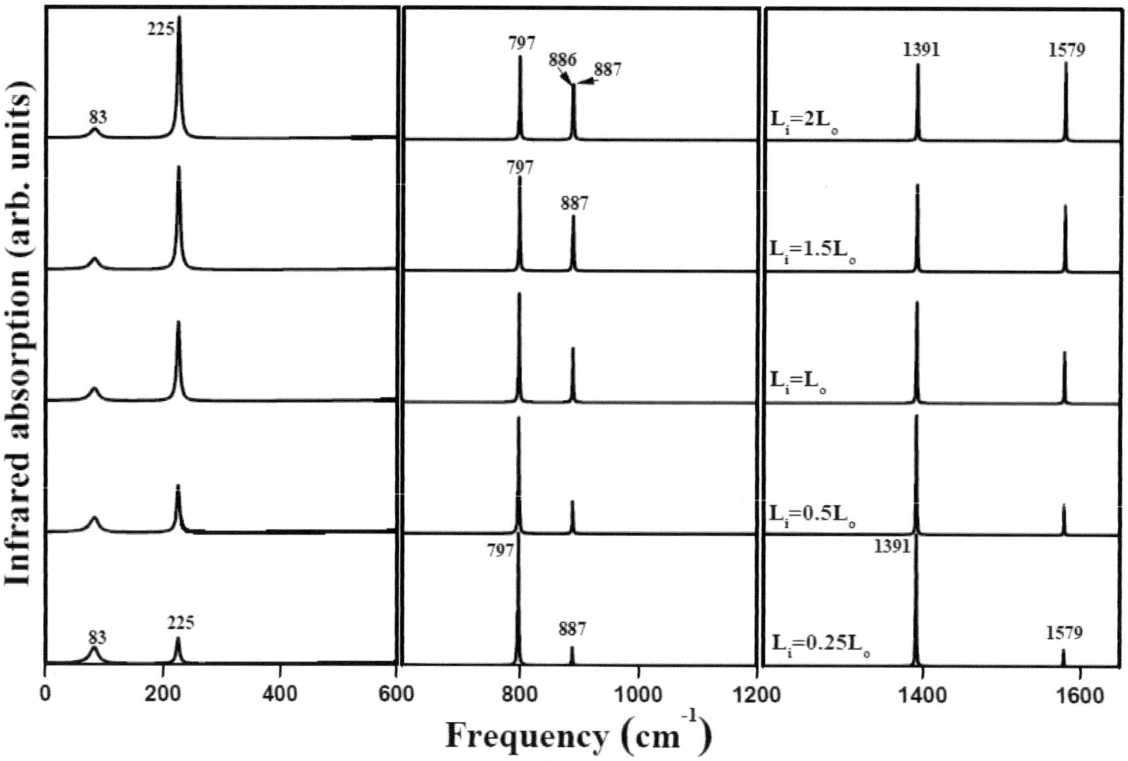

Figure 6. Calculated $I_x(\omega)$ infrared spectra of the (9,0)@(18,0) CBN as a function of the L_i /L_o ratio.

In the x-polarization, and for usual CBN nanotubes (L_i=L_o), in the TLM region the infrared spectrum is dominated by two bands located at 1391 and 1579 cm^{-1} in the (9,0)@(18,0) CBN nanotubes. For L_i > L_o, the intensity of mode at 1579 cm^{-1} increases and is assigned to the TM of the part of the inner carbon tube, which does not interact with the outer boron nitride tube (the TM of the corresponding (9,0) SWCNT is calculated at 1580 cm^{-1}). For Li < Lo, the intensity of mode at 1391 cm^{-1} increases and is assigned to the TM of the part of the outer boron nitride tube that does not interact with the inner carbon tube (the TM of the corresponding (18,0) SWBNNT is calculated at 1391 cm^{-1}). BLM and ILM range shows the same dependence of the Li/Lo ratio than the TLM range and the above discussion is also valid. The intensity of the lines at 83 and 797 cm^{-1} (respectively 225 and 887 cm^{-1}) decreases when the Li/Lo ratio increases (respectively decreases).

4. Conclusion

We have outlined the construction of double wall nanotube consisted of carbon nanotube inside BN nanotube. For this obtained system, the dependence of the infrared spectra as a function of the diameter and finite lengths of outer and inner nanotubes have been analyzed using the spectral moments method. First, we calculated the infrared spectra of infinite length of CBN nanotubes as a function of their diameters, we observed that the mode located in BLM region decrease when the diameter increase. We also calculated the infrared spectra of CBN nanotubes with a finite length. We observed additional infrared bands as a function of the relative lengths of the inner and outer tubes in the whole infrared spectrum. By varying the length of the CBN sub-system only bands located in the BLM and ILM regions have changed.

References

[1] Chopra N G, Luyken R J, Cherrey K, Crespi V H, Cohen M L, Louie S G and Zettl A 1995 *Science.* **266** 966-67

[2] Iijima S 1991 *Nature.* **354** 56-58

[3] Chang C W et al 2006 *Phys. Rev. Lett.* **97** 085901

[4] Stewart D A, Savic I, Mingo N 2009 *Nano Lett.* **9** 81–84

[5] Chen Y et al 2004 *Appl. Phys. Lett.* **84** 2430–32

[6] Blase X, Rubio A G, Louie S and Cohen M L 1994 *Europhys. Lett.* **28** 335

[7] Zhi C et al 2009 *Adv. Func. Mater.* **19** 1857–62

[8] Lauret J S et al 2005 *Phys. Rev. Lett.* **94** 037405

[9] Mickelson W, Aloni S, Han W Q, Cumings J and Zettl A 2003 *Science.* **300** 467–69

[10] Trave A, Ribeiro F J, Louie S G, Cohen M L 2004 *Phys Rev B.* **70** 205418

[11] Walker K, Rance G A, Pekker A, Tohati H M, Nemeth G, Fay M W, Lodge R W, Stoppiello C T, Kamaras K and Khlobystov A N 2017 *Small Methods.* **1** 1700184

[12] Lee C H, Wang J, Kayatsha V K, Huang J Y and Yap Y K 2008 *Nanotechnology.* **19** 455605

[13] Guo L and Singh R N 2009 *Phys. E.* **41** 448

[14] Lim S H, Luo J, Wei J and Lin J 2007 *Catalysis Today.* **120** 346-50

[15] Bi J Q, Wang W L, Qi Y X, Bai Y J, Pang L L, Zhu H L, Zhao Y and Wang Y 2009 Mater. Lett. B. 63 1299-1302

[16] Fakrach B, Rahmani A, Chadli H, Sbai K, Bentaleb M, Bantignies J L and Sauvajol J L 2012 *Phys. Rev. B.* **85** 115437

[17] Bantignies J L, Sauvajol J L, Rahmani A and Flahaut E 2006 *Phys. Rev. B.* **74** 195425

[18] Rahmani A, Sauvajol J L, Rolsand S and Benoit C 2002 *Phys. Rev. B.* **66** 125404-12

[19] Fakrach B, Fergani F, Boutahir M, Rahmani AH, Chadli H, Hermet P and Rahmani A 2018 *Crystals.* **8** 118

[20] Darkrim F and Levesque D 1998 *J. Chem. Phys.* **19** 4981-84

[21] Lee C H, Wang J, Kayatsha V K, Huang J Y and Yap Y K 2008 *Nanotechnology.* **19** 455605

IOP Conf. Series: Materials Science and Engineering **783** (2020) 012010 doi:10.1088/1757-899X/783/1/012010

[22] Yuan J and Liew K M 2011 *Carbon.* **49** 677–683

[23] Saito R, Takeya T, Kimura T, Dresselhaus G, Dresselhaus M S 1998 *Phys. Rev. B.* **57** 4145

[24] Benoit C, Royer E and Poussigue G 1992 *J Phys. Condens Matter.* **4** 3125

IOP Conf. Series: Materials Science and Engineering **783** (2020) 012011 doi:10.1088/1757-899X/783/1/012011

CFD Simulations and experimental investigation of nucleate pool boiling of liquid nitrogen

Kenza Bouazaoui[1*], Rachid Agounoun[1], Imad Kadiri[1] and Khalid Sbai[1]

[1]Université Moulay Ismail (UMI), Laboratoire d'Etude des Matériaux Avancés et Applications (LEM2A), Ecole Supérieure de Technologie de Meknès, Km 5, route d'Agouray, N6, 50040, Morocco.

*Corresponding author: Email: k.bouazaoui@edu.umi.ac.ma

Abstract. Pool boiling heat transfer is a very efficient mode of heat transfer. It is used in various energy conversion and heat exchange systems and in the cooling of high energy density electronic components. In this work, heat transfer during nucleate boiling for a brass ribbon horizontally immersed in liquid nitrogen is experimentally and numerically studied. An experimental apparatus was built to conduct a pool boiling heat transfer study. The capabilities of Computational fluid dynamics (CFD) boiling model in pool boiling heat transfer is investigated. The computational model used combines the Euler/Euler two-phase flow. The general measurements of the super-heat, the influences of super-heat on the triggering of the boiling, as well as the critical flux which represents an important issue for the security of the systems were analyzed. The results of numerical simulations were compared with experimental data.

1. Introduction

Boiling heat transfer has always been a very interesting field of research because it is the most efficient way to remove the heat from a solid body with liquids and to reach the highest heat fluxes at the minimum wall superheat. Several studies [1, 2, 3] have focused on understanding fluid-solid combinations in order to obtain the highest heat fluxes with minimal superheating of the wall. Consequently, the process of boiling heat transfer has an important place in the industrial sector, particularly in the fields of electric power generation and refrigeration. Due to the complex physics of boiling phenomenon, it is studied intensively by experimental analyzes and the numerical simulations. In the first branch, several experimental studies are carried out to the process of heat transfer in boiling and which have proven to be effective on the analysis of the characteristics of boiling in transient state and in the stationary state [4-6].

The numerical simulations study of the boiling process also proven to be capable of reliably predicting phase-change (liquid/vapor) operation and heat transfer characteristics.

In recent years, Computational Fluid Dynamics (CFD) codes are considered to be the most promising numerical tools used for the analysis boiling heat transfer. Particularly, in the case of boiling flows, where heat is transferred into the fluid from a heated wall at such rates that boiling and vapor is generated.

Generally, most open-source CFD codes correspond to phase change are adapted to a multiphase Eulerian approach based on a two-fluid model. In the case of boiling heat transfer, where heat is transferred into the fluid from a heated wall, the multiphase CFD boiling models are usually coupled to appropriate wall boiling sub-models due to additional source terms describing the boiling physics at the heated wall that must be included in the equations [7]. Yi Liu et al [8] developed the CFD (Computational Fluid Dynamics) code to investigate the pool boiling of a cryogenic liquid. The model included the wall temperature effect on the vaporization velocity of the cryogenic liquid. The different boiling regimes (boiling nucleation, transition zone and boiling in film) were simulated by the use of a single model. The heat flux calculated for each boiling regime was compared with experimental data from the literature. This heat flux, which depends on the temperature difference between the wall and the cryogenic liquid, is different for each stage of boiling [9]. The CFD code was chosen to examine the complexity of the boiling phenomenon. In this sense, Tang et al [10] explored numerically the boiling mechanism of cryogenic fluids use of CFD model. Using a whole heating section, including boiling surface (polished with 5000# emery paper), which was immersed in LN2 pool, and the pool is inside a vacuum Dewar (with vacuum degree of the order of 10-4 Pa).The simulation focused on analysis the nucleate boiling process of the bubbles formation at the exit of the heated wall and the influences of super-heat on bubble departure diameters. The efficient cooling of the heating surface is limited by an important factor such as the critical flux (CHF). The latter appears when the relative rate between vapor and liquid reaches a critical value producing instability of the interface (Helmholtz instability) and preventing rewetting of the wall [11]. The security of the system is based on the determination of CHF. More different models developed for the prediction of maximum thermal flux and that are based on classical theories of critical flux [12, 13]. Recently, the most efficient determination of critical flux has been performed by numerical models well developed. For example, the numerical study of Lal et al focused on the prediction of CHF within the framework of Computational Fluid Dynamics (CFD) with an interface tracking (IT) method. In this paper, we present the simulation of nucleate heat boiling on a brass ribbon horizontally immersed in liquid nitrogen. The numerical simulation model proposed for this investigation is based on a CFD boiling model and the heat transfer model. The wall boiling model is developed in Euler-Euler multiphase frame work that solves transport equations for each phase. The model present in this paper is developed for analyzed the all boiling curve of LN2 pool boiling (natural convection area and nucleate boiling area) in atmospheric pressure.

2. Numerical model setup

Improving knowledge of the physical mechanisms governing boiling and the development of CFD have stimulated the construction of phenomenological models to study boiling heat transfer phenomena. These models can then be integrated into commercial calculation codes.
In this section, the configuration of the boiling system, the basic equations and the equations of the boundary conditions used in COMSOL for numerical simulations are recalled. The main steps of a typical model are then presented.

2.1. Configuration of the boiling system

Figure 1. Shown the configuration of domain used in solver COMSOL multi-physics. The configuration is a 2D model, it is presents a heater element (brass ribbon) immersed in a container filled with liquid nitrogen (LN2). The immersed cell has very fine dimensions (length = 10^{-1} m, width = 2.10^{-3} m, thickness = 50.10^{-6} m). The brass ribbon is heated by a heat flux density between q= 2.40 W/cm², q= 13.8 W/cm².The latter shows the heater surface for realizing the nucleate boiling phenomenon.

IOP Conf. Series: Materials Science and Engineering **783** (2020) 012011 doi:10.1088/1757-899X/783/1/012011

3. CFD model

In COMSOL multi-physics [15], the wall boiling models are developed in the context the multiphase Euler-Euler model. The CFD model used is basised of a model contains the equations of motion (Navier–Stokes equations) and the equation of energy conservation.

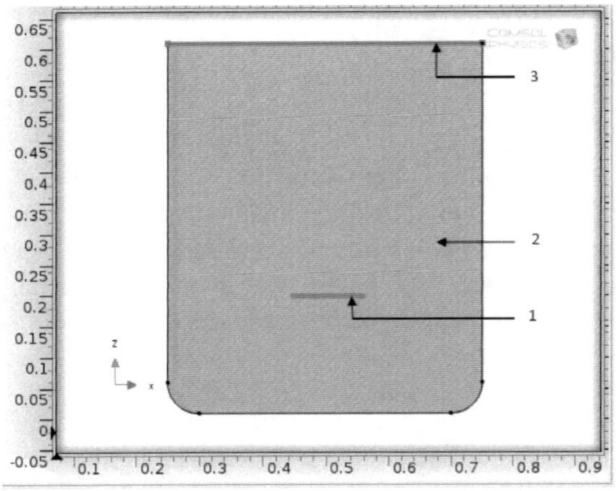

Figure 1. The computation domain of the boiling system, (1) Brass ribbon, (2) Liquid nitrogen, (3) Free surface

Continuity equations:

$$\frac{\partial \alpha_l \rho_l}{\partial t} + \nabla.(\alpha_l \rho_l u_l) = 0 \tag{1}$$

Transport equation for the volume fraction of the vapor phase:

$$\frac{\partial \alpha_v \rho_v}{\partial t} + \nabla.\ \alpha_v \rho_v u_v = m_{l \to v} \tag{2}$$

Where $m_{l \to v}$ presented the mass transfer rate from liquid to vapor (kg/ (m^3·s)). The subscripts “l” and “v” denote quantities related to the liquid phase and the vapor phase, respectively.

Momentum equation:

$$\alpha\rho \frac{\partial u}{\partial t} + \alpha\rho(u.\nabla)u = \nabla.\ \left[-p + \alpha\mu(\nabla u + (\nabla u)^T \right] + \alpha\rho g + F \tag{3}$$

Energy equation:

$$\rho\ C_p \frac{\partial T}{\partial t} + \rho\ C_p u\ .\nabla T = \nabla.\ (k\nabla T) + Q \tag{4}$$

Energy equation for the heated wall:

In the wall, the heat transfer model uses the heat equation or heat conduction equation as the mathematical model for heat transfer in solid. The heat conduction equation is:

$$\rho \ C_p \frac{\partial T}{\partial t} = \nabla . \ (k \nabla T) \tag{5}$$

3.1. Boiling model

In the previous works [16, 17] of boiling heat transfer, it is possible to account the phase change model. The boiling heat transfer rate is necessarily related to the phase change rate, the mass transfer mainly controls the heat transfer. This mass transfer rate is typically depends on the interfacial area between liquid and vapor phases. The model takes care of the phase change near the wall as well as inside the domain. The mass transfer from liquid to vapor during evaporation is described as following [29]:

$$m_{l \to v} = C \rho_l \frac{T - T_{sat}}{T_{sat}} \quad T \succ T_{sat} \tag{6}$$

Where C is the mass transfer intensity factor or relaxation time (s^{-1}) in the vapor-liquid phase change [18].
The following table summarizes the input parameters used in the numerical model developed.

Table 1. Thermal properties of the materials

Material properties	Brass	Nitrogen	Unit
Density (ρ)	8386	808,4	$kg.m^{-3}$
Heat capacity (C_p)	377	1040	$J.kg^{-1}.k^{-1}$
Thermal conductivity	121	145.10^{-3}	$w.m^{-1}.k^{-1}$
Specific heat ratio (γ)	1,3967	2,07	1
Dynamic viscosity (μ)	-	$5,435.10^{-6}$	Pa.s

4. Experimental setup

4.1. Description

To validate the numerical results, an experimental study was performed. The experimental device is composed of brass ribbon, power supply system, Dewar tank of liquid nitrogen (LN2), nitrogen liquid, and data acquisition system, as shown in Fig.2. The brass ribbon is immersed permanently in a Dewar tank filled with liquid nitrogen (LN2). Furthermore, a platinum probe Pt-100 used for measured the

fluid temperature in boiling vessel. Finally, a camera for the observation of boiling complete the device. The experiments are performed under atmospheric pressure (1atm).
In our experiment study the variation of heat flux density q versus superheat ΔT (boiling curve) is determined from the heat flux density and the wall superheat. The accuracy on the determination of the temperature and flux density is 0.3 K and 0.1 W / cm², respectively.
The heat flux dissipated by Joule effect and the average ribbon superheat ΔT are determined as follows:

$$\dot{q} = \frac{U_{ribbon} \cdot I_{in}}{S_e} \quad (7)$$

$$\Delta T = T - T_{LN2} \quad (8)$$

T presents the wall temperature; this temperature is determined by the electrical resistivity ρ Eq.9

$$\rho(T) = \frac{R S_t}{L} = \frac{U S_t}{I L} = \rho_0 (1 + \sigma_0 T) \quad (9)$$

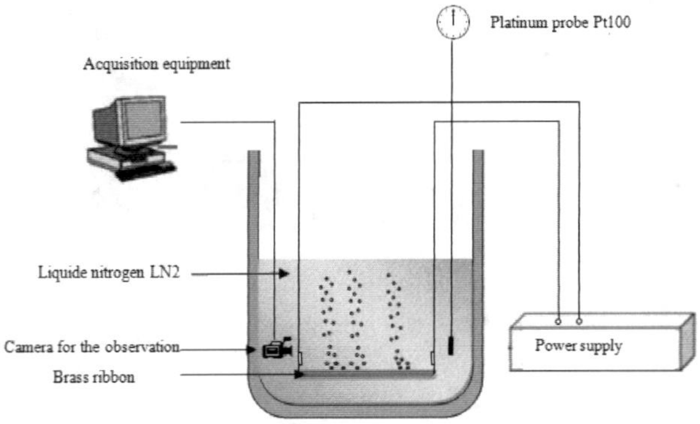

Figure2. Experimental setup composed of container of liquid nitrogen(LN2), brass ribbon, acquisition system and power supply system.

5. Result and discussion

5.1. Simulation result

After the monophasic exchange phase (natural convection phase), boiling is produced when the element heater temperature reaches a maximum.
Fig. 3 shows the numerical results for the volume fraction distribution of vapor at nucleate boiling in the heater element surface(brass ribbon). A typical burst of small masses of vapor has been presented at the ribbon level. The existence of vapor on the surface has just explained the effect of heater wall on the liquid nitrogen evaporation. Indeed, when the wall is heated, a heat quantity transmitted by the heating element surface to the liquid, which is the liquid nitrogen near the heated wall, changed its liquid state to a set of vapor bubbles on the area. More particularly, the heat quantity generated by the

heating element surface in the boiling liquid increases with the increase of the wall temperature. This is confirmed by the Nukiyama study [19] with a metal surface immersed in water as a coolant.
The small bubbles of vapor appeared on the heated surface in isolation for a medium heat flux density. The increase in the imposed heat flux density allows to increase the quantity of vapor on the heated surface.

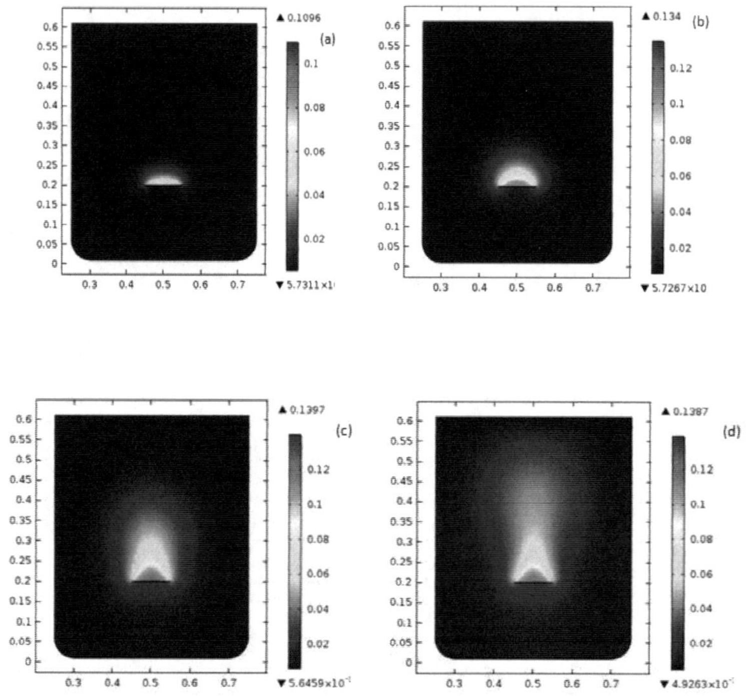

Figure 3. Simulated nucleate pool boiling on a brass ribbon plunged in liquid nitrogen.

5.2. Boiling curve

Figure 4 shows the results obtained in steady state regime of a brass ribbon mounted in horizontal position and immersed in a Dewar tank filled with liquid nitrogen (LN2). The manipulations and simulations allowing us to find the boiling curve (Nukiyama curve[11,19]). They highlight 3 zones: the natural convection regime, the nucleate boiling regime and the film boiling regime. The pure conduction regime can not be visualized because this regime exists for very small imposed flux density values.
The boiling curve of LN2 is begins by a natural convection area for low flux density. In this area, vapor has not occurred on the surface. The bubbles begin to appear for an average flux density of 1.2 W/cm2 and a superheat of 11.3 K. the increase in imposed flux density makes it possible to increase the number of bubbles or the quantity of vapor on the heating surface.
When the flux density is imposed, it corresponds to the limit value of flux density (critical heat flux CHF) that can be exchanged without there abrupt rise in the heating element temperature and risk of deterioration. The CHF or maximum flux is approached at an average flux density of 16.2W/cm2 and a superheat of 11.2 K.
The simulated boiling curve is also presented in Fig.4. The experiment boiling curve between heat flux and super-heat for LN2 is used for comparison to evaluate our CFD simulations. Similarly, we note

that the low heat flux between 0,30 W/cm² and 1,2W/cm² corresponding to natural convection regime. As the heat flux increases, the super-heat increases accordingly during the nucleate boiling process. In this regime, superheat in around 7.86 to 10.52K, and the applied heat flux ranges from 4.77 to 16.2W/cm². The vapor emerged from the solid surface are clear.
Simulated values were compared with the experimental results with a maximum error of 2,7% for pool boiling conditions investigated in this work.

6. Conclusion
In this paper, the LN2 boiling curve between heat flux and super-heat for experiment and CFD simulation study is presented and analysed. CFD formulations based on Euler-Euler multiphase model and the heat transfer model was built up. The simulations are performed in COMSOL Multiphysics software. The brass ribbon is used for the heating element. The liquid nitrogen is chosen as the working fluid due to its distinctive properties in hot wall cooling. Simulations were mainly realized for different heat fluxes applied and under atmospheric conditions. The results show the boiling curve regimes (natural convection regime and nucleate boiling regime) corresponding to low and high values of heat flux applied. The numerical simulation between heat flux and super-heat for LN2 pool boiling were in good accordance with the measured data.

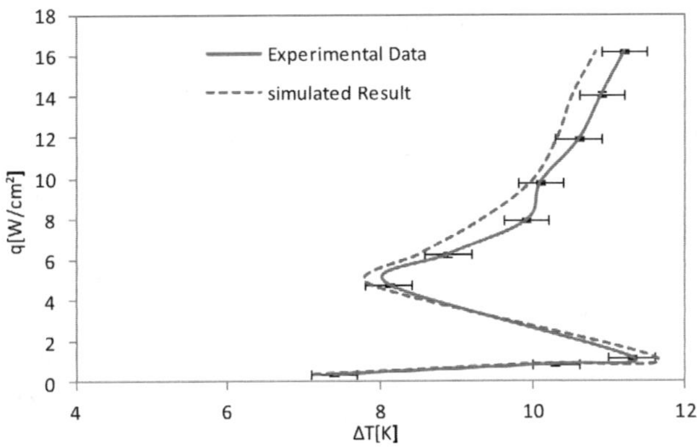

Figure 4. LN2 pool boiling curve: Experimental data and CFD simulation results.

References
[1] Zhang B J, Kim K J, 2014, Nucleate pool boiling heat transfer augmentation on hydrophobic self-assembly mono-layered aluminia nano-porous surfaces, *Int. J. Heat Mass Transf,* **Vol .73** p 551-561 .

[2] Xu P, Li, Xua Y, 2015 Enhanced boiling heat transfer on composite porous surface, *Int. J. Heat Mass Transf,* **Vol .80** p 107 114 .

[3] Sanna A, Hutter C, Kenning D B R, Karayannis T G, Sefiane K and Nelson R A, 2014 Numerical investigation of nucleate boiling heat transfer on thin substrates, *Int. J. Heat Mass Transf,* **Vol .76** p 45-64 .

[4] Héas S, Launay S, Raynaud and Lallemand M, 1998, Transient nucleate boiling heat transfer from a thick flat sample. Proceedings of the second International Symposium on Two-Phase Flow modeling and experimentation, *Pise, Ed: E.T.S,* **Vol. I**, p 205-210.

[5] Drach V, Fricke J, 1996, Transient heat transfer from smooth surfaces into liquid nitrogen. *Cryogénies*, **Vol. 36**, p. 263-269.

[6] Duluc M C, Francois X, 1998, Steady-state transition boiling on thin wires in liquid nitrogen. The role of Taylor wavelength, *Cryogenics* **Vol .38** p 631-638.

[7] Kurul N, Podowski M Z, 1990, Multidimensional effects in forced convection subcooled boiling, Proc. *of the 9th International Heat Transfer Conference*, Jerusalem, Isral.

[8] Liu Y, Olewski T, Vechot L, Gao X, Mannan S, 2015, Modeling of a cryogenic liquid pool boiling by CFD simulation, *Journal of Loss Prevention in the Process Industries* **Vol .35**, p125-134 doi:10.1016/j.jlp.2015.04.006.

[9] Barron R F, 1999, *Cryogenic Heat Transfer*. Taylor & Francis, Philadelphia, USA.

[10] Xiaobin Z , Wei X, Jianye C, Yuchen W and Tang K, 2015, CFD simulations and experimental verification on nucleate pool boiling of liquid nitrogen, *Physics Procedia* **Vol .67**,p 569 – 575 , doi:10.1016/j.phpro.2015.06.077

[11] Lallemand M, 1991, Ebullition en vase. Entropie, N° 160, *Universités à l'Institut National des Sciences Appliquées de Lyon*, BE 8 235, p. 4-19.

[12] Zuber N, Hydrodynamic aspects of boiling heat transfer. *PhD. Thesis, University of California,* Los Angeles, CA, USA (also published as USAEC. doi 10.2172/4175511.

[13] Haramura Y, Katto Y 1983, A new hydrodynamic model of critical heat flux, applicable widely to both pool and forced convection boiling on submerged bodies in saturated liquids. *Int. J. Heat Mass Transfer*, **Vol. 26**, N° 3, p.389-399.doi:10.1016/0017-9310(83)90043-1

[14] COMSOL Multiphusics, COMSOL 4.3a, May 2012 http://www.comsol.com.

[15] Punekar H, Das S 2013, Numerical Simulation Jacket of Subcooled Nucleate Boiling in Cooling, *IC Engine, SAE International* , doi:10.4271/2013-01-1651.

[16] Liub J, Wanga G, Zhanga L, Y.Shia, Zhangb H andYao S 2017, Numerical simulation of single bubble boiling behavior, *Propulsion and Power Research* 6(2), p 117–125.

[17] Tian Y, Zhang K, Wang N, Cui Z and Cheng L 2017, Numerical study of pool boiling heat transfer in a large-scale confined space, *Appl. Therm. Eng* 118, p 188–198.

[18] Nukiyama S 1966, The maximum and minimum values of the heat Q transmitted from metal to boiling water under atmospheric pressure, *Int. J. Heat Mass Transf.* **Vol .9**, p1419–1433.

IOP Conf. Series: Materials Science and Engineering **783** (2020) 012012 doi:10.1088/1757-899X/783/1/012012

Infrared-Active Modes in Finite and Infinite Hexagonal Boron Nitride

Oussama Boutahir[1], Souhail Lakhlifi[1], Mourad Boutahir*[1], Brahim Fakrach[2], Abdelhai Rahmani[1], Hassane Chadli[1], Abdelali Rahmani[1]

[1]Laboratoire d'Etude des Matériaux Avancés et Applications, Université Moulay Ismail, Faculté des Sciences Meknes, Morocco
[2]Laboratoire de Physique Théorique et Appliquée, Université Sidi Mohammed Ben Abdellah, Faculté des Sciences Dhar El Mahraz Fez, Morocco
***Corresponding author e-mail: mourad.boutahir@gmail.com**

Abstract. In this theoretical paper, we study the Infrared spectra of h-BN as a function of their lengths. Calculations are performed in the framework of the force constants model, using the spectral moments method. This original approach allows us to consider not only infinite h-BN sheet as usual in most theoretical models but also h-BN sheet with finite lengths. These predictions are useful to interpret the experimental data.

1. Introduction

Hexagonal boron nitride (h-BN) is an interesting quasi two-dimensional insulator [1]. It has attracted intense scientific and engineering interest due to its excellent lubricity, high thermal, chemical stability and especially remarkable electronic properties. The band structure engineering of h-BN extends the exploitation of its potential applications.

Atomically thin two-dimensional (2D) materials [1,2] such as graphene, hexagonal boron nitride (BN), and the transition metal dichalcogenides (TMDCs) offer a variety of outstanding properties for fundamental studies and applications. More recently, the capability to assemble multiple 2D materials with complementary properties into layered heterogeneous structures presents an exciting new opportunity in materials design [2–8], but several fundamental challenges remain, including making good electrical contact to the encapsulated 2D layers. Electrically interfacing 3D metal electrodes to 2D materials is inherently problematic. For graphene devices, the customary approach is to metalize the 2D surface. In multilayer structures, the requirement to expose the surface for metallization presents additional restrictive demands on the fabrication process. For example, encapsulated BN/graphene/BN heterostructures (BN-G-BN) need to be assembled sequentially so as to leave the graphene surface accessible during metallization, because no process to selectively remove the BN layers has been identified. In this paper we focus on the infrared-active modes of BN sheet and BN nanoribbons

2. Model, Method & Approach:

The spectral moments method [9] was shown to be a powerful tool for determining vibrational spectra (infrared absorption, Raman scattering and inelastic neutron-scattering spectra) of harmonic systems.

IOP Conf. Series: Materials Science and Engineering **783** (2020) 012012 doi:10.1088/1757-899X/783/1/012012

This method can be applied to very large systems, whatever the type of atomic forces, the spatial dimension, and structure of the material.

For boron nitride intratube interactions of h-BN, we used the same force constant as used in recent calculations of the Infrared spectra of SWBNNTs [10]. The intensity of the infrared absorption spectrum is given by the following expression:

$$I_\alpha(\omega) = \frac{\omega}{nc}\sum_j \frac{|a_{j\alpha}|^2}{2\omega_j}\left(\delta(\omega - \omega_j) - \delta(\omega + \omega_j)\right) \quad (1)$$

With

$$a_{j\alpha} = \sum_{k\alpha} \frac{q_k}{\sqrt{m_k}} e_j(k\alpha) \quad (2)$$

Where n and c are respectively the index of the material and speed of light, respectively, ω_j and e_j(kα) are respectively the frequency and the (kα) component of the displacement amplitude for the kth (α is cartesian coordinate) in the jth mode. q_k is the effective charge and m_k is the mass of the kth atom. To enhance the IR response of nanotubes, dynamical effective charges on tied carbon atoms were fixed at q_A=+1 and q_B=1 for a given A-B bond. The usual method to calculate the IR spectrum consists of injecting in the previous expressions the values of ω_j and e_j(kα) obtained by direct diagonalization of the dynamical matrix of the system. However, when the system contains a large number of atoms, as for long dimer of SWNT, the dynamical matrix is very large and its diagonalization fails or require long computing time. By contrast, the spectral moment's method [9] allows the Infrared spectrum of very large harmonic systems to be directly computed without any diagonalization of the dynamical matrix

3. Results & Discussion:

In this section, we report the calculated infrared spectra for infinite BN sheet, which is obtained by applying periodic conditions. In order to identify all IR active-modes and their symmetries, we present in figure 1 the ZZ calculated Infrared spectra of BN sheet. Spectrum is displayed in the Intermediated wave-number (IWN) and high-wave-number (HWN) ranges.

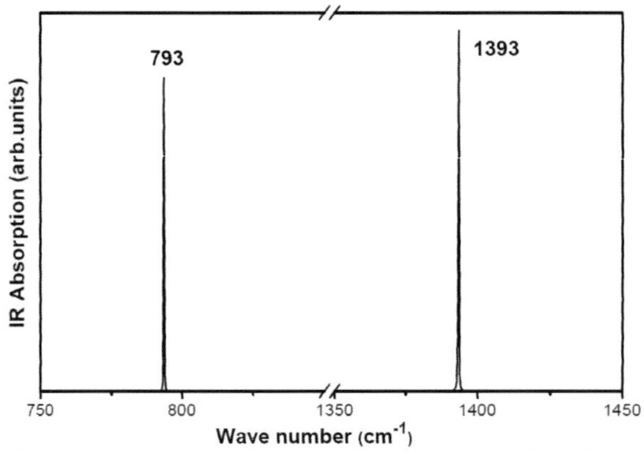

Figure 1. The ZZ polarized infrared spectra calculated for a h-BN sheet. Spectra are displayed in the Intermediated-like modes (ILM) and tangential-like modes (TLM) ranges.

Concerning the intermediate-wave-number (IWN) region, from 600 to 1200 cm^{-1}, the spectrum is dominated by a strong single mode located around 793 cm^{-1} of A_{1u} (A1) symmetry. This mode corresponds to an out-of-plane mode displacements of B and N atoms in the opposite directions, as illustrated in figure2.

In the high-wave-number (HWN) range, above 1200 cm^{-1}, the ZZ polarized spectra are dominated by strong IR-active modes. The ZZ spectra evidence that the Au mode is located at 1393 cm^{-1}. The eigenvector displacements of this mode are given in figure2, (the arrows indicate the magnitude and direction of B and N atom displacements). These vectors allow us to identify the longitudinal or transverse character of the optical modes. In the HWN range, all the modes are tangential modes.

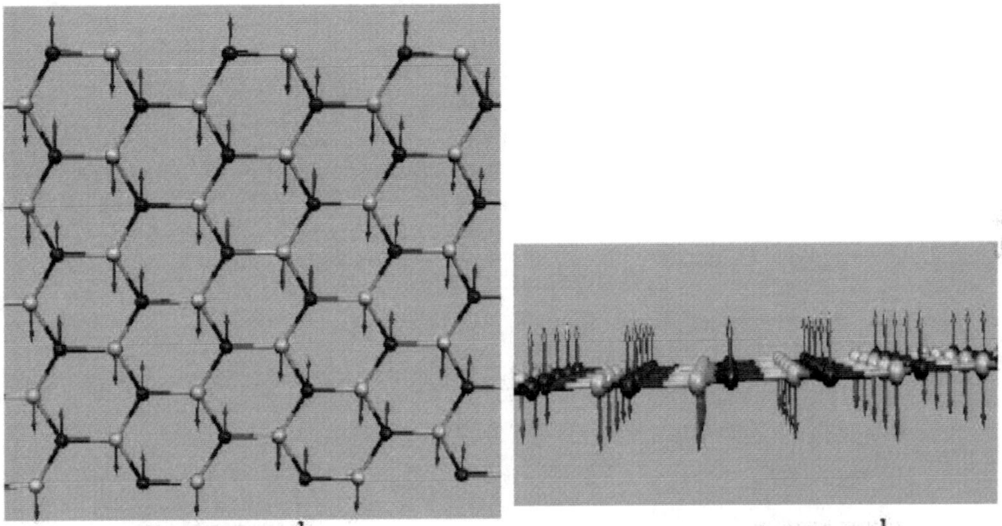

Figure 2. Infrared-active-mode atomic displacements and symmetries for selected normal modes in the h-BN sheet.

The most important advantage of the spectral moment's method is the possibility to calculate the infrared spectra for a finite and long width of BN nanoribbon. Now, we focus on Dependence of the width of nanoribbon. We calculated the ZZ infrared spectra of h-BN as a function of the width in the ILM (left) and TLM (right) see figure3. Spectra are displayed for five finite widths, L = 2.15, 4.30, 8.61, 21.52, 86.10 nm, and for infinite h-BN, from bottom to top

From the figure 3 we show the ZZ polarized spectra of BN nanoribbon. We can see the existence of many active modes spread over two zones of the spectrum. Starting with the high wave-number region, one can clearly see the main mode detected around 1393 cm^{-1} that correspond to A_{1u} (A1) symmetry, and other new peaks appearing through edge effects. We next explore IWN region, also we can clearly see the main mode detected around 793 cm^{-1}. Several other peaks are observed.

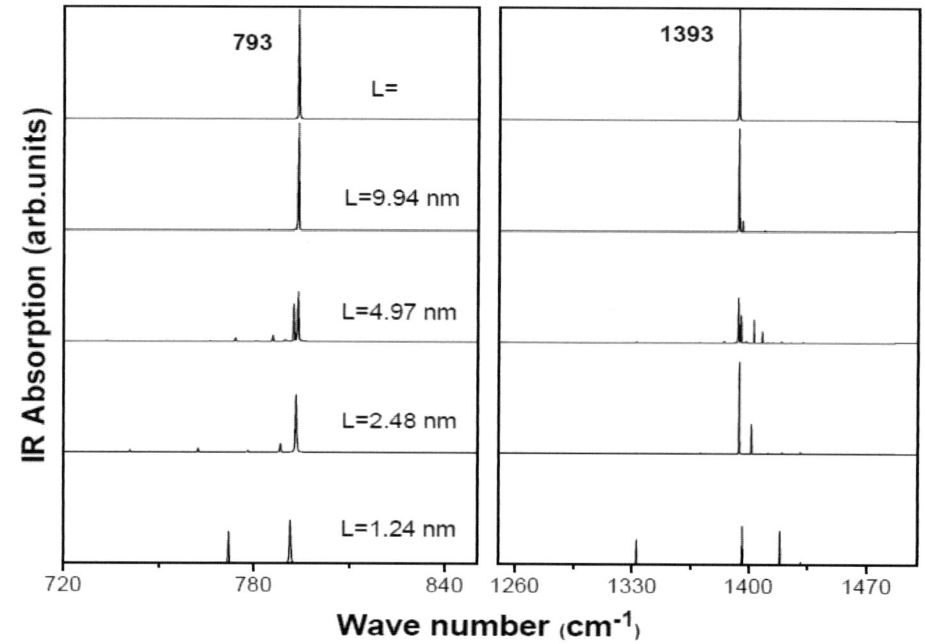

Figure 3. Dependence of the ZZ in the ILM (left) and TLM (right) infrared spectra of h-BN as a function of the length. Spectra are displayed for five finite lengths, L = 2.15, 4.30, 8.61, 21.52, 86.10 nm, and for infinite h-BN, from bottom to top.

In this paragraph, we have reported in figure4 the calculated electronic band structures of h-BN monolayer. One can see that in the case of the monolayer, all spectral features remain essentially the same as for bulk h-BN, with the exception of the gap between the valence band (VB) and conduction band (CB), which has a significantly higher value (5 eV).

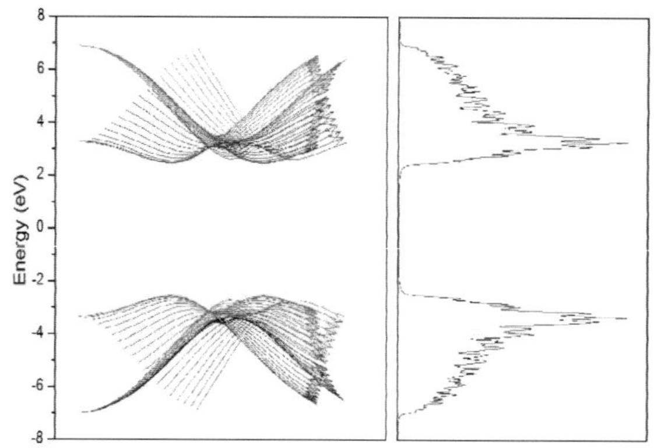

Figure 4. The calculated electronic band structures (left) and density of state (right) of h-BN monolayer

4 Conclusion

In this work, we have calculated the Infrared spectra in the ILM and TLM ranges of h-BN, using the spectral moment's method. The dependence of these modes has been calculated as a function of their chirality and lengths. The finite-size effects are shown to be significant only for short lengths, and the main effect of the length shortening is the appearance of additional peaks in the intermediate and tangential infrared domain.

References

[1] Xu M, Liang T, Shi M and Chen H 2013 *Chem. Rev.* **113** 3766–98
[2] Geim A K, Grigorieva I V 2013 *Nature.* **499** 419–425
[3] Mayorov A S et al 2011 *Nano Lett.* **11** 2396–99
[4] Dean C R et al 2013 *Nature* **497** 598–602
[5] Ponomarenko L A et al 2013 *Nature.* **497** 594–597
[6] Georgiou T et al 2013 *Nat. Nanotechnol.* **8** 100–103
[7] De Lillo F, Cecconi F, Lacorata and Vulpiani A 2008 *EPL.* 84 40005
[8] Topsakal M, Aktürk E and Ciraci S 2009 *Phys. Rev. B.* 79 115442
[9] Rahmani A, Sauvajol J L, Rols S and Benoit C 2002 *Phys. Rev. B.* **66** 125404
[10] Fakrach B, Rahmani A, Chadli H, Sbai K and Bentaleb M 2012 *Phys. Rev. B.* **85** 115437

Effect of the integration of hemp wool as an insulation material for the construction of the roof and external walls of a typical Moroccan building

M.Dlimi[1,*], R.Agounoun[1], I.Kadiri[1], K.Sbai[1] and A.Rahmani[2]

[1]Laboratoire d'Etude des Matériaux Avancés et Applications (LEM2A), Université Moulay Ismail (UMI), Ecole Supérieure de Technologie de Meknès, Km 5, route d'Agouray, N6, 50040, Meknes, Morocco.
[2]Laboratoire d'Etude des Matériaux Avancés et Applications (LEM2A), Université Moulay Ismail (UMI), Faculté des Sciences de Meknès, bp 11201, Avenue Zitoune, Meknes, Morocco.

[*]Corresponding author : m.dlimi@edu.umi.ac.ma

Abstract. This paper deals with the integration of thermal insulation in Moroccan buildings roofs and external walls in order to reduce energy consumption due to the use of air conditioning systems. The insulation material under investigation is the hemp wool, an eco-friendly material whose use allows the storage of the Carbone and help reducing greenhouse gas emissions. In order to assess the dynamic thermal performance of the typical Moroccan insulated building, a life cycle cost analysis is conducted based on annual cooling transmission loads specific to the climate of Meknes, Morocco. Optimum insulation thicknesses, energy savings and payback periods are determined for a building life-span of 20 years. Investigations are carried out based on an implicit finite difference method using a home-made code developed in Matlab.
Keywords: energy consumption; energy savings; hemp wool; optimum insulation thickness; payback period.

1. Introduction

In the context of sustainable development, the new thermal regulations in the building sector are orienting consumers towards the use of bio-based insulation materials. Contributing to the reduction of greenhouse gas emissions and temporary carbon storage, these renewable materials are derived from either plant biomass or animal biomass.

Thanks to its porous structure and its thermal insulating properties, hemp is increasingly used in eco-construction. It can be used as hemp concrete inside a wood frame structure, or as hemp wool for insulation, or even as hemp-lime mortar for coating. Colinart et al. [1] conducted an experimental study to assess the hygrothermal behavior of three hemp concrete walls. The first one was uncoated, and the others were coated with lime-sand and hemp-lime plasters. Results showed that, when the wall is coated, the levels of vapour pressure are more dampened compared to the uncoated wall, and the moisture transfers are modified thanks to the low permeability of hemp-lime and sand-lime plasters.

Costantine et al. [2] investigated experimentally the thermal behavior of hemp concrete for external building wall insulation. Results obtained for indoor temperature and relative humidity underlined the ability of hemp concrete to dampen the outdoor climatic conditions. Then, the SPARK environment was used to validate the experimental approach. Good agreement was shown between experimental measurements and numerical values. The thermal performance of insulation materials was assessed by Gori et al [3]. The authors studied the effect of thermal insulation position on the thermal performance of a bi-layered wall. Obtained results showed that the most effective position of the insulating layer is when it is located on the outside of the building structure. Furthermore, in order to decrease the energy consumption and to realize maximum energy savings, insulation materials must be used with optimum insulation thickness. Indeed, Nematchoua et al. [4] calculated the required insulation thickness with the corresponding energy savings for buildings in two different cities with two different climates. Two wall structures were considered: compressed stabilized earth wall and concrete block wall. The insulation material used was the extruded polystyrene. Results showed that in the city of Yaounde which is characterized by an equatorial climate, optimum insulation thickness and energy savings for the south facing wall were equal to 0.08 m and 51.69 $/m², respectively. For the city of Garoua which is characterized by a tropical climate, optimum insulation thickness calculated for a north orientation wall was obtained as 0.11 m and energy savings as 97.82 $/m². Furthermore, Based on the isotropic model of Duffie and Beckman [5], Ibrahim et al. [6] determined the optimum insulation thickness according to cooling requirements in buildings located in the city of Beirut, Lebanon. The insulation material chosen was extruded polystyrene, and the optimum insulation thicknesses obtained are 4.9, 4.8, 4.1 and 4.1 for the west, east, north and south orientations, respectively.

In this study, hemp wool is used as bio-based insulation material. Afterwards, considering only cooling requirements specific to the Moroccan city of Meknes, a life cycle cost analysis is applied in order to calculate optimum insulation thicknesses for roof and external walls of a Moroccan typical building.

2. Analysis and modelling

2.1. Studied configurations

(a)

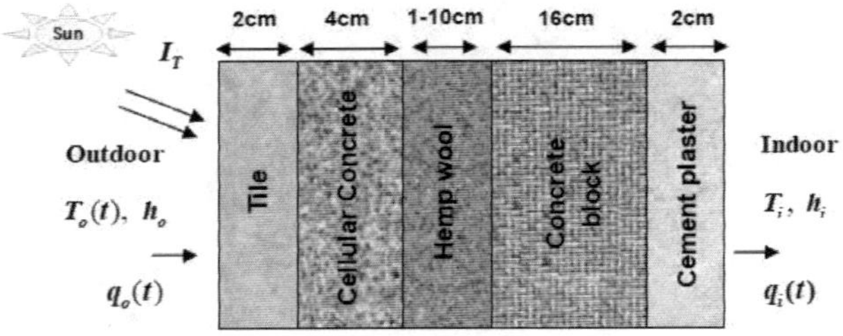

(b)

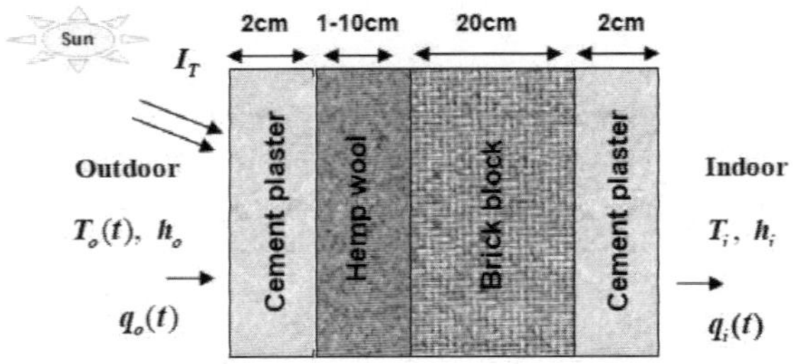

Figure 1. (a): Roof configuration, (b): External wall configuration

Two composite wall structures are considered: the first one concerns a Moroccan typical roof configuration and the second one describes a typical external wall configuration. In the two wall structures, an insulation layer varying from 1 to 10 cm is introduced and located on the external side of the multilayer walls.

The insulation material chosen in our study is the hemp wool, a bio-based insulation material with thermo-physical properties close to those of the expanded polystyrene, the most classical insulation material [7].

The materials used in the wall structure and their thermal properties are given in Table 1.

Table 1. Thermo-physical properties of building materials [8] [9] [10].

Material	k (W/m.K)	ρ (kg/m^3)	C_p (J/kg.K)
Brick block	0.62	1800	840
Concrete block	0.51	1400	1000
Cellular Concrete	0.22	600	880
Hemp wool	0.04	35	1200
Tile	1.704	2300	700
Cement plaster	0.72	1865	840

2.2. *Mathematical model*

Considering constant thermophysical properties and no heat generation, the heat transfer through each layer of the external wall is governed by the transient 1D heat equation expressed as:

$$k_j \frac{\partial^2 T_j}{\partial x^2} = \rho_j c_j \frac{\partial T_j}{\partial t}, \quad j = 1,2,...,M \tag{1}$$

where k_j, ρ_j and c_j are the thermal conductivity, the density and the specific heat of the jth layer, respectively. T_j is the temperature, x and t are spatial and temporal coordinates.

Eq.1 is solved with the specification of an initial and two boundary conditions. As initial condition, an arbitrary uniform temperature field is assumed. The outdoor and indoor boundary conditions are given as follows, respectively [11]:

$$-k_1 (\frac{\partial T}{\partial x})_{x=0} = h_o (T_o(t) - T_{x=0}) \tag{2}$$

$$-k_M (\frac{\partial T}{\partial x})_{x=L} = h_i (T_{x=L} - T_i) \tag{3}$$

where $h_i = 9W/m^2 \cdot K$ and $h_o = 22W/m^2 \cdot K$ are the combined inner and outer heat transfer coefficients. T_i is the indoor air temperature maintained constant at a fixed design temperature $T_i = 26°C$ according to Moroccan Standard NM ISO 7730 RTCM [12]. The outside surface of the wall is exposed to the sol-air temperature $T_o(t)$ which includes solar radiation effects on the outdoor temperature and is expressed as follows [11]:

$$T_o = T_a + \frac{aI_T}{h_o} - \frac{\varepsilon \Delta R}{h_o} \tag{4}$$

where T_a is the outdoor air temperature. a and I_T are the external wall surface solar absorptivity and the total solar radiation, respectively. $\frac{\varepsilon\Delta R}{h_o}$ express the correction factor and is assumed to be 4°C for horizontal surfaces and 0 for vertical ones [13].

The outdoor air temperature T_a and the hourly values of total solar radiation I_T are computed using the commercial thermal simulation software TRNSYS [14] based on Meteonorm data specific to the city of Meknes (Latitude: 33°89' N, Longitude: 5°54' W). The external wall surface solar absorptivity is taken equal to 0.55 [15].

The implicit finite difference method [16] is applied to solve the transient 1D heat conduction equation.

Equations are solved using matrix functions in Matlab [17]. The accuracy of the implicit finite-difference method is assessed by checking the model and making sure that its results are grid and time-step independent.

2.3. *Economic model for the insulation thickness optimization*

The optimum economic insulation thickness represents the value providing the lowest total life-cycle cost.

Using cooling transmission loads as input data, and based on the present worth method, an economic model is set for the determination of the optimal insulation thickness.

Cooling transmission loads are calculated from June to September. The day 15 of each month is considered as a representative day for the whole month.

The hourly variation of the inside surface heat flux is expressed as follows [18]:

$$q_i = h_i(T_{x=L} - T_i) \tag{5}$$

By integrating the instantaneous inside surface heat flux over a 24h period, daily total loads are calculated and then, yearly cooling loads are obtained by the addition of daily loads over the summer period considered from June to September.

Afterwards, the cost of energy consumption during a building lifetime of N years is given by [19]:

$$C_{enr} = PWF\left(\frac{Q_c}{COP}\frac{C_{el}}{(3.6\times10^6)}\right) \tag{6}$$

PWF is the present worth factor, Q_c, C_{el} and COP represent the annual cooling transmission loads per unit area, the cost of electricity for cooling and the cooling system performance, respectively.

The total cost is the sum of the cost of energy consumption and the cost of insulation material. The total cost per unit area of wall is given by:

$$C_t = C_{enr} + C_i = PWF\left(\frac{Q_c}{COP}\frac{C_{el}}{(3.6\times10^6)}\right) + C_{ins}L_{ins} \tag{7}$$

C_{ins} is the cost of the insulation material by unit volume and L_{ins} is the insulation thickness.

Annual energy savings A_s are obtained by the division of the difference between the energy costs of the wall without insulation and the wall with optimum insulation thickness by the present worth factor PWF.

The payback period P_b is defined as the insulation cost divided by annual energy savings, and is calculated as follows:

$$P_b = \frac{C_i \cdot L_{ins(opt)}}{A_s} \tag{8}$$

All the parameters needed for the above economic calculations are summarized in Table 2.

Table 2. Parameters used in calculations.

Parameter	Value
Electricity cost, C_{el} [20]	0.1346 \$/kWh
Coefficient of performance, COP [21]	2.5
Insulation cost, C_{ins} [22]	100 \$/m^3
Interest rate, g [23]	2.25 %
Inflation rate, i [23]	1.8 %
Lifetime period, N [24]	20 years

3. Results and discussion

The optimum insulation thickness is determined based on a life cycle cost analysis using cooling transmission loads as its main inputs.

The variation of the energy consumption cost, the insulation cost and the total cost in respect to insulation thickness for the two configurations studied are depicted in figures 2 and 3.

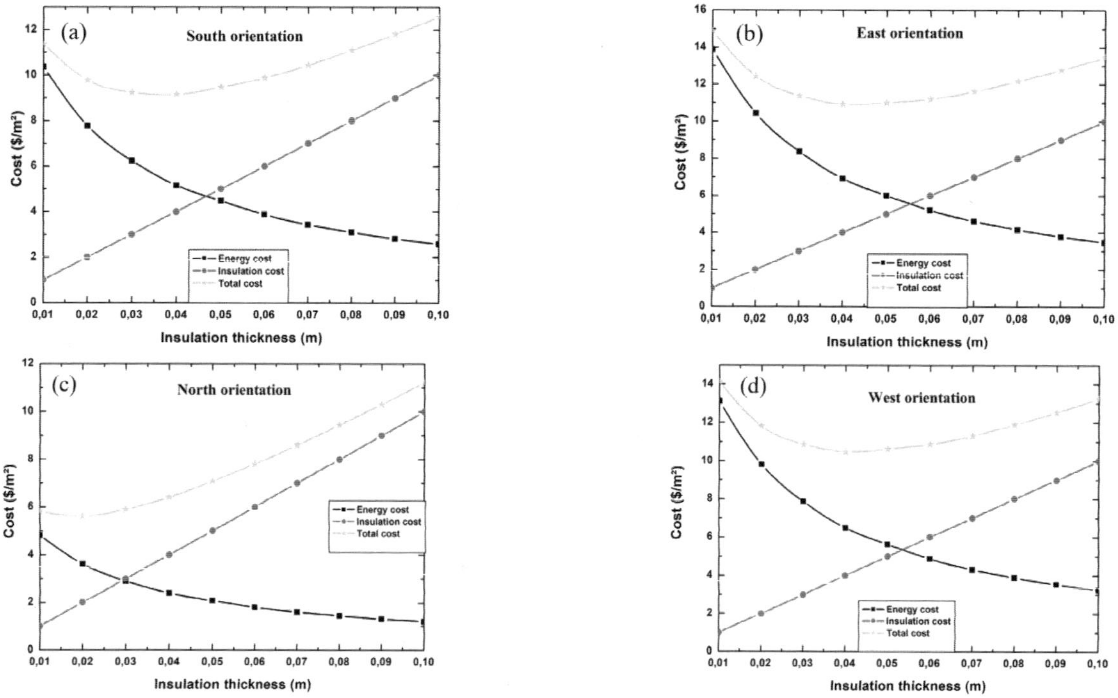

Figure 2. Costs variation with insulation thickness for the external wall structure: (a): south orientation, (b): east orientation, (c): north orientation and (d): west orientation.

It is clearly shown from figures that the energy cost decreases, while the insulation cost knows a linear increase with the increase of the insulation thickness. The total cost, which represents the sum of energy and insulation costs, allow the determination of the optimum insulation thickness. It corresponds to the lowest value shown by the total cost curve.

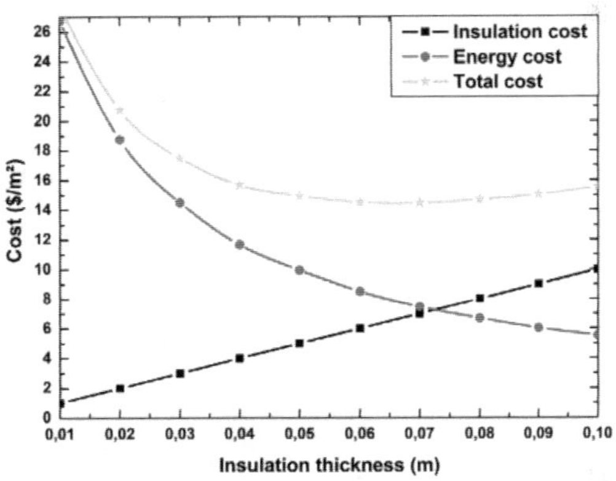

Figure 3. Costs variations with insulation thickness for the roof structure.

For the roof configuration, the optimum insulation thickness is obtained as 7 cm, whereas for the external wall configuration, optimum insulation thicknesses are obtained as 2 cm for the north orientation and 4 cm for south, east and west orientations.

Table 2 summarizes the values of optimum insulation thicknesses, minimum total costs, energy savings and payback periods for the roof and all orientations of the external wall.

Table 3. Results summary.

Wall orientation	Optimum insulation thickness (m)	Minimum total cost ($/m²)	Energy savings ($/m²)	Payback period (years)
Roof	0.07	14.4658	18.1104	3.0266
South	0.04	9.1539	6.6251	7.1528
North	0.02	5.6154	1.7205	20.42
East	0.04	10.9222	10.2275	5.3417
West	0.04	10.4861	9.7949	5.5092

One can note that the minimum and maximum total costs are provided respectively by, the northern external wall and the roof structure.

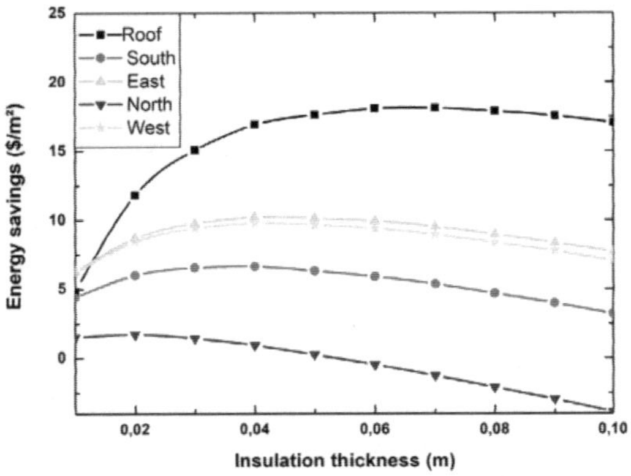

Figure 4. Energy savings for roof and external wall.

However, in terms of energy savings, the roof insulation achieves energy savings of 63.41, 43.52, 90.5 and 45.91 % more than south, east, north and west orientations, respectively (Figure 4).

Regarding the payback period, the smallest value is provided by the roof structure, followed by the east, west, south and north facing external walls.

Finally, as presented in figure 5, the use of hemp wool with optimal thicknesses for the thermal insulation of a Moroccan building has reduced annual air conditioning requirements by 77% for the roof. For the external walls, this reduction reached approximately 68% for the south, east and west orientations, and 50% for the north orientation.

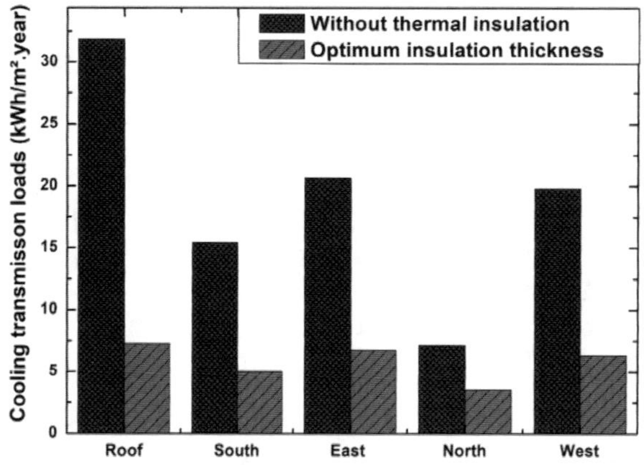

Figure 5. Annual cooling transmission loads required by roof and external walls.

4. Conclusion

In this study, the dynamic thermal performance of a Moroccan typical building whose roof and external walls are insulated with hemp wool is investigated. The building is exposed to the dynamic thermal conditions specific to the city of Meknes.

Firstly, a computer program was developed in Matlab and a finite difference method was used to provide a numerical solution of the transient heat transfer equation through a multilayer wall.

Afterwards, considering a building lifespan of 20 years and based on a life-cycle cost analysis, optimum insulation thickness, energy savings and payback period were determined through the development of an economic model where the yearly cooling transmission loads are used as main inputs. Optimum insulation thicknesses are obtained as 7 cm for the roof, 4 cm for south, east, and west orientations, and 2 cm for the north orientation.

The lowest optimum insulation thickness is obtained by the north facing wall since it provides the lowest annual cooling loads. It can be considered as the most economical case. Nevertheless, this orientation gives negligible energy savings and very high payback period.

Finally, a comparison between an uninsulated building and an insulated one showed that annual cooling requirements can be reduced by at least 77 and 65% respectively, when the roof and external walls are insulated with optimum thicknesses of hemp wool.

References

[1] Colinart T, Glouannec P, Pierre T, Chauvelon P and Magueresse A 2013 *Buildings* **3** 79–99.
[2] Costantine G, Maalouf C, Moussa T and Polidori G 2018 *Build Environ* **131** 140–153.
[3] Gori P, Guattari C, Evangelisti L and Asdrubali F 2016 *Int. J. Heat Mass Tran.* **103** 349–359.

[4] Kameni Nematchoua M, Ricciardi P, Reiter S and Yvon A 2017 *Int. J. Sustain. Environment.* **6** 170–182.
[5] Duffie J.A and Beckman W.A 1991 Solar Engineering of Thermal Processes, New York: John Wiley and Sons, Inc.
[6] Ibrahim M, Ghaddar N and Ghali K 2012 *J. Build. Perform. Simul.* **5** 384–398.
[7] Liu L, Li H, Lazzaretto A, Manente G and Tong C 2017 *Renew. Sust. Energ. Rev.* **69** 912–932.
[8] Ozel M 2013 *Energy Convers. Manag.* **66** 106–114.
[9] Jin X, Zhang X, Cao Y and Wang G 2012 *Energ. Buildings.* **47** 369–374.
[10] Guechchati R, Moussaoui M.A, and Mezrhab A 2012 *Appl. Sol. Energy.* **48** 33–37.
[11] Threlkeld J.L 1998 *Thermal Environmental Engineering*, Englewood Cliffs, NJ: Prentice-Hall.
[12] Règlement Thermique de Construction au Maroc (RTCM), *Agence Nationale pour le Développement des Energies Renouvelables et de l'Efficacité Energétique (ADEREE)*.
[13] American Society of Heating, Refrigerating and Air-Conditioning Engineers, ASHRAE Standard: Standards for Natural and Mechanical Ventilation. New York: The Society. (1973).
[14] TRNSYS17, Transient System Simulation Program, *Solar Energy Laboratory, University of Wisconsin-Madison, WI, USA.*
[15] Reagan J.A and Acklam D.M 1979 *Energ. Buildings.* **2** 237–248.
[16] Ozisik M.N 1994 Finite Difference Methods in Heat Transfer, *Boca Raton, Florida: CRC Press.*
[17] MATLAB, MATrix LABoratory - 7.10.0. User's Guide, MathWorks, 2010.
[18] Cengel Y, Heat transfer: a practical approach 1ed ed.: *Hightstown: McGrawHill,* 1998.
[19] Daouas N 2011 *Appl. Energy* **88** 156–164.
[20] RADEM (Régie Autonome de Distribution Eau et Electricité).
[21] Ozel M 2012 *Energ. Buildings* **49** 552–559.
[22] Guide Technique sur l'Isolation Thermique du Bâtiment au Maroc, Agence Nationale pour le Développement des Energies Renouvelables et de l'Efficacité Energétique (ADEREE).
[23] BANK AL-MAGHRIB (Central Bank of Morocco).
[24] Ozel M 2011 *Appl. Therm. Eng.* **31** 3854–63.

IOP Conf. Series: Materials Science and Engineering **783** (2020) 012014 doi:10.1088/1757-899X/783/1/012014

Raman spectroscopy analysis of single wall carbon nanotubes with penta- and hexa-vacancies defects

SA Ait Abdelkader, M Boutahir, AH Rahmani, B Fakrach, M Bentaleb, H Chadli and A Rahmani

Advanced Material and Applications Laboratory (LEM2A), University Moulay Ismail,FSM-ESTM-FPE, BP 11201, Zitoune, 50000 Meknes, Morocco.

E-mail: a.rahmani@fs-umi.ac.ma

Abstract. In this paper the effect of penta- and hexa-vacancies defects on a vibrational properties of single wall carbon nanotube (SWNT) was investigated. The bond-polarizability model combined with the spectral moments method are carried out to calculate the non-resonant Raman spectra. The changes of Raman spectra as a function of the nanotube chirality and the vacancies defects concentration on the wall of the nanotube are identified. The intensity ratio of D-band to G-band (I_D/I_G ratio) reflects the purity of SWNTs. Our study offers a means to distinguish between pristine and defective SWNTs even at low defect concentrations.

1. Introduction

Since the discovery of carbon nanotubes (CNT) [1], due to their various and remarkable properties [2, 3, 4, 5], several efforts have been made regarding their potentials applications [6, 7, 8, 9], and much attention has been devoted to the investigation of their properties [10, 11, 12, 13]. While a relatively simple description of the nanotube properties can be achieved by using the perfect CNT structure. However, in real CNTs, defects are unavoidable in non-equilibrium conditions, e.g., during the nanotube growth [14, 15, 16]. Various kinds of defects, such as impurities [18, 17], Stone-Wales (SW) defects [20, 19] and vacancies; like the mono-vacancy (MV) [21, 22, 23], di-vacancy (DV) [24, 25, 26] and its derivative point defect can be contained in the CNTs samples.

The mechanical, optical and electrical properties of carbon nanotubes (CNTs) and related materials have been an active research fields [1, 27, 28, 29]. In contrast to theoretical considerations the experimental verifications of the CNTs strength or Young's modulus demonstrate evidently discrepancies that may reach even up to 30% [30]. Structural defects in nanotubes have been identified by STM and HRTEM imaging [31, 32]. Possible single or multiple defects in CNTs provide an explanation for the extant theoretical experimental differences. While such defects in single-wall carbon nanotubes (CNTs) have a great influence on their physical properties [33, 34]. Then, the evaluation of effects of defects and impurities on properties of these nano-structures would be crucial because of the inevitable creation of defects during the synthesis and purification processes [35, 36] and under mechanical strains [37]. Among structural defects, vacancy and Stone-Wales defects [38] are more prevalent and their numbers and locations would be unclear and it is necessary to investigate their influences from different points of view.

CNTs characterizations relies extensively on Raman spectroscopy [34], it's one of the most powerful tools to the analysis of CNTs of varying diameter and chirality [39]. In the Raman spectra of defective CNTs and all sp^2 hybridized disordered carbon materials, the defect-induced phonon mode so-called D band is observed [40, 41] at around 1350 cm^{-1} [42, 43, 44]. It's a signature of the presence of defects or disorder in the graphene lattice [45], and is often used to characterize defective CNTs. The D band intensity can be enhanced as function of the number of defects is increased in CNTs. The D band has been used for the assessment of imperfection of SWNTs and the understanding of the properties of their defects. However, it is possible that D band is derived from a many defective CNTs in the sample, while the radial breathing mode (RBM) or graphitic G-band are due to the presence of the other CNTs in the same sample, then it may be the interpretation of the D-band feature might be ambiguous. Further, identifying Raman modes associated with points defects can lead the Raman spectroscopy to a more effective tool for the characterization of defects signatures.

The purpose of the current study is to reveal the dependence on the polarized ZZ Raman spectra of penta- and hexa-vacancy defect in the single wall carbon nanotube as a function of defect concentration and chirality of nanotube. Also we calculate the ratio of D- to the G-band intensity (I_D/I_G) which provides a parameter that can be used to quantifying the defects. The non-resonant Raman spectra are calculated in the framework of spectral moments method [10], together with a bond polarizability model.

2. Model and method

The key element of this study is Raman spectroscopy, such as a perfect tool to get important information and characterization of both perfect and defective carbon nanotube and peapod. In this work, we focus on vibrational properties of penta- and hexa-vacancies defect in SWNT (Figure 1). For this purpose we described the C-C intratube interactions at the surface of SWNT by using the force constant model introduced by Wirtz and Rubio [46], but for the interaction of carbon atoms sited in the vacancy defect border are calculated by using Density Functional Theory (DFT) as implemented inside the SIESTA package [47].

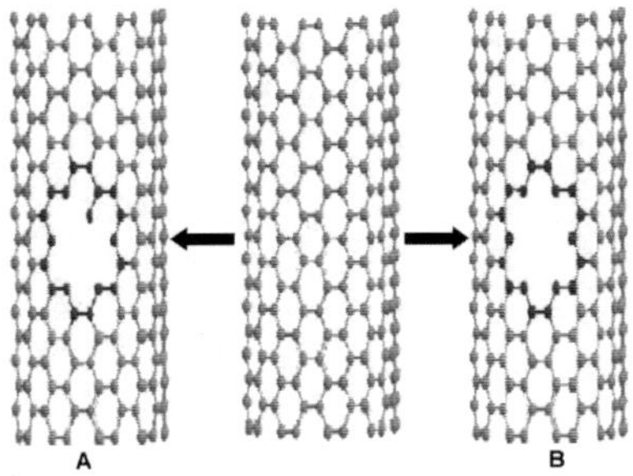

Figure 1. Molecular models of (A) penta-vacancy and (B) hexa-vacancy defect.

The Raman intensity is calculated according to the bond polarizability model [48, 49]. In this approach, the polarization is only modulated by the nearest-neighbor bonds and the components of the induced polarizability tensor $\tilde{\pi}$ are given by the empiric equation:

$$\pi_{\alpha\beta}(r) = \frac{1}{3}(\alpha_l + 2\alpha_p)\delta_{\alpha\beta} + (\alpha_l - \alpha_p)\left(\hat{r_\alpha}\hat{r_\beta} - \frac{1}{3}\delta_{\alpha\beta}\right) \quad (1)$$

where α and β relate to the Cartesian components (x,y,z) and $\vec{\hat{r}}$ is the unit vector along the vector $\vec{r}$ connecting atom n and atom m which are covalently bonded. The parameters α_l and

α_p correspond to the longitudinal and perpendicular bond polarizability respectively. Within this approach, one can assume that the bond polarizability parameters are functions of the bond lengths r only. The derivatives $\pi^n_{\alpha\beta,\gamma}$ are given by:

$$\pi^n_{\alpha\beta,\gamma} = \sum_m \frac{1}{3}(\alpha'_l + 2\alpha'_p)\delta_{\alpha\beta}\hat{r_\gamma} + (\alpha'_l - \alpha'_p)\left(\hat{r_\alpha}\hat{r_\beta} - \frac{1}{3}\delta_{\alpha\beta}\right)\hat{r_\gamma} + \frac{\alpha_l - \alpha_p}{r}(\delta_{\alpha\gamma}\hat{r_\beta} + \delta_{\beta\gamma}\hat{r_\alpha} - 2\hat{r_\alpha}\hat{r_\beta}\hat{r_\gamma}) \quad (2)$$

where $\alpha' = \left(\frac{\partial\alpha}{\partial r}\right)\Big|_{r=r_0}$ and r_0 is the equilibrium bond distance. The bond polarizability parameters are:

$$\overline{\alpha} = 2\alpha'_p + \alpha'_l, \quad \overline{\beta} = \alpha'_l + \alpha'_p, \quad and \quad \overline{\gamma} = \frac{\alpha_l - \alpha_p}{r} \quad (3)$$

The values of theses parameters are fitted to reproduce the experimental Raman spectrum.

3. Results and discussion

In this section a complete investigation between perfect and penta- (hexa-) vacancies defects sited in the wall of SWNT is conducted. In this study, the concentration of defects is defined by the number of the removed atoms devided by the total number of atoms in the perfect tube. Our systems are large nanotubes composed by 80000 carbon atoms, whose the maximum number of penta- (Hexa-) vacancies defects can reach 640 (533) respectively. In all our Raman spectra calculations, the nanotube axis is along the Z-axis and a carbon atom is along the X-axis of the nanotube reference frame. The laser beam is kept along the Y-axis of the reference frame. We consider that both incident and scattered polarizations are along the Z-axis to calculate the polarized ZZ spectra.

3.1. Raman Spectra of carbon nanotubes with penta-vacancies defects

We study the Raman spectra of defective SWNT as a function of defect concentration and chirality. For this purpose, three chiralities of SWNTs have been considered with a diameter of 1.356 nm [(10,10) armchair], 1.335 nm [(14,5) chiral], and 1.330 nm [(17,0) zigzag] at a penta-vacancies concentration varied from 2% to 4%. The calculated ZZ-polarized Raman spectra of penta-vacancy nanotube are shown in the panels of Figure 2 along with their corresponding pure nanotubes. The spectra are displayed in the RBM range (left), in intermediate range (middle), and in the range of tangential modes (right).

It is found from Figure 2 that compared with that in the perfect SWNTs, although the penta-vacancies leads to a redshift of the tube's RBM frequency. When the concentration of penta-vacancies increases, RBM downshift from 164 cm^{-1} [166 and 167 cm^{-1}] in pristine tube to 161 cm^{-1} [164 and 165 cm^{-1}], 158 cm^{-1} [161 and 162 cm^{-1}], and 156 cm^{-1} [158 and 159 cm^{-1}] for 2%, 3%, and 4% penta-vacancies concentration in (10,10) [(14,5) and (17,0)] SWNT, respectively. A second-lowest Raman-active peak is observed in the case of defect tube only and located at 340 cm^{-1} [330 and 336 cm^{-1}], 338 cm^{-1} [328 and 334 cm^{-1}], 334 cm^{-1} [325 and 330 cm^{-1}] for 2%, 3%, 4% penta-vacancy concentrations for (10,10) [(14,5) and (17,0)] nanotube, respectively.

In the intermediate range, the spectrum is dominated by a strong bands at around 1342 cm^{-1} in three chiralities of SWNT, while no Raman line is expected for perfect SWNT. This mode arises from penta-vacancy defect and its position slightly dependent on the tube chirality and on defect concentration. Similarly to the previous calculations for SWNTs, the G-band

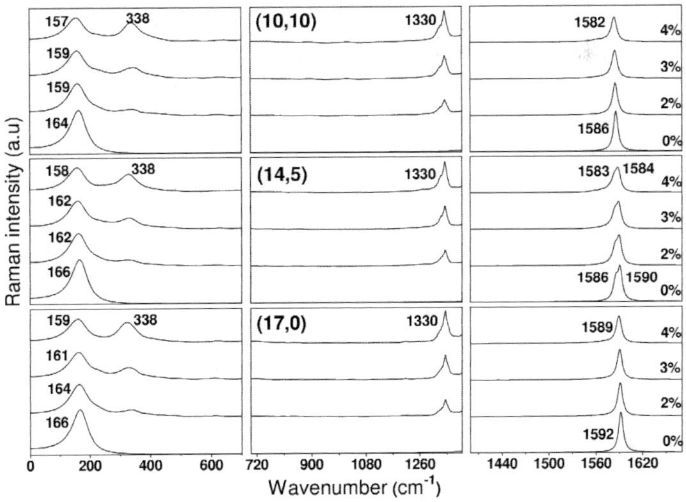

Figure 2. ZZ-polarized Raman spectra of different SWNT chirality as a function of penta-vacancies defects concentration.

located in the TM range depends on the tube chirality, is weakly downshift as a function of defect concentration. For instance, the G mode downshifts from 1586 cm^{-1} in the perfect (10,10) nanotube to 1582 cm^{-1} in the 4% defective nanotube. When the concentration of penta-vacancies increases this mode weakly downshifts for the three chiralities.

The eigenvector displacements of the modes located around 338 cm^{-1} and 1342 cm^{-1} in (10,10) SWNT with penta-vacancy defect, are given in Figure (3-A). These vectors allow us to identify the radial displacements of atoms belonging to the penta-membered ring: all atoms in the penta-vacancy move in phase in the radial direction and the mode is fully symmetric. This mode could be used to estimate penta-vacancy defect of nanotubes.

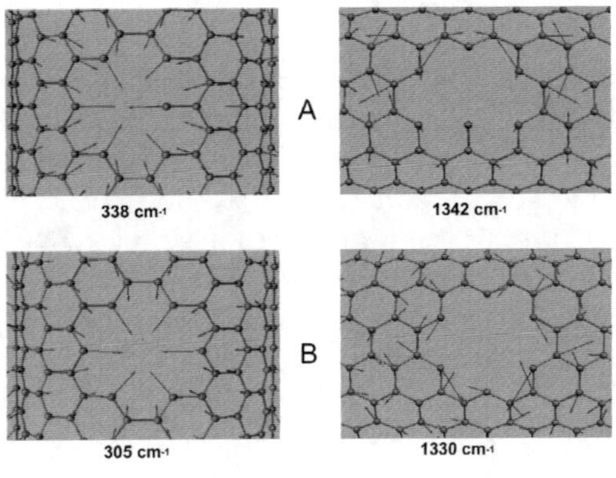

Figure 3. A : Raman active mode atomic displacements for 338 cm^{-1} and 1342 cm^{-1} (resp. 305 cm^{-1} and 1330 cm^{-1}) normal modes in the (10,10) nanotube with penta-vacancy defects (resp. B : hexa-vacancy defects).

In the present study, a systematic investigation of the relationship between the Raman spectroscopic response and microstructural changes of CNTs were performed. In practice, the use of Raman spectroscopy to study the structural quality of pristine and defective SWNT is limited to the evaluation of the I_D/I_G ratio, where I_D is the intensity of the D band, activated by the presence of defect in carbon systems, and I_G is the intensity of the G band. In fact, the decrease of the intensity of the G mode upon defect creation increases the I_D/I_G intensity ratio, which could lead to an estimation of the amount of defects.

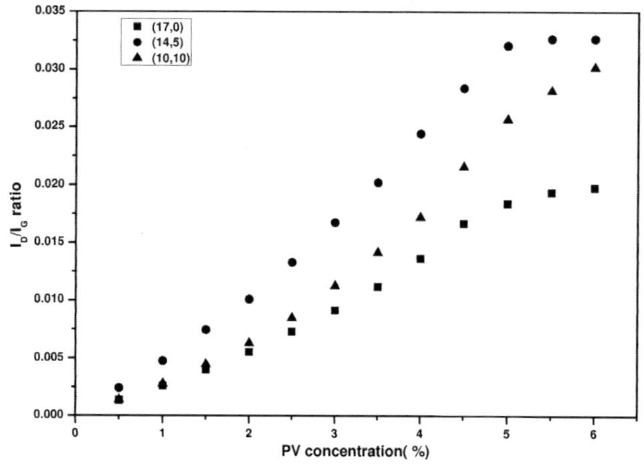

Figure 4. The evolution of the intensity ratio I_D/I_G as a function of the penta-vacancies defects concentration in armchair (triangle), zigzag (square), and chiral (circle) nanotubes.

Figure 4 shows the normalized I_D/I_G intensity ratio as a function of defect concentration (DC) for three chiralities of SWNT (armchair (10,10), zigzag (17,0) and chiral (14,5)). From 0 to 5.5% concentration, the I_D/I_G ratio increased linearly until it peaked at 0.0325 [0.0196] for (14,5) [(17,0)] penta-vacancy tube. After this value of DC, the I_D/I_G intensity ratio is considerably stable, it is clear that the main discrepancy between both behavior of (14,5) and (17,0) defect tube is caused by the difference chirality between them. Contrariwise, the armchair defect tube behavior increased without reaching its stable state.

3.2. Raman Spectra of carbon nanotubes with hexa-vacancies defects

The figure 5 shows the calculated ZZ polarized Raman spectra of perfect SWNT and hexa-vacancy defect SWNT. Three concentration of hexa-vacancy defect are considered (2%, 3%, 4%), and the results are compared with perfect one: (top) armchair ((10,10) tube diameter close of 1.356 nm), (middle) chiral ((14,5) tube diameter close of 1.335 nm), and (bottom) zigzag nanotubes ((17,0) tube diameter close of 1.330 nm). The ZZ Raman spectra can be divided into three selected frequency regions: the RBM range (left), the region of the defect modes (middle), and tangential range: G-modes of the tubes (right).

By examining the Raman active modes calculated for defective nanotube compared with the perfect one, as shown in figure 5, a notable effect in the defective SWNTs is the reduction of the intensity of active modes independently of the nanotube chirality compared to the pristine SWNT. This result can be understood due to the loss of symmetry induced by the defects. Additionally, in the high-frequency region, the G-mode red-shifted when the concentration of defect increase independent of chirality of nanotube. For instance the G-mode frequency passed from 1586 cm^{-1} in perfect nanotube to 1582 cm^{-1} in the highest concentration of hexa-vacancy defect in (10,10) SWNT. As for intermediate range between 600 and 1400 cm^{-1}, a new mode is showed around of 1330 cm^{-1}. This mode located in the D-band of SWNTs Raman spectra arises from hexa-vacancy defect, whose is not sensitive to the defect concentrations and not to SWNT chirality. Within the 0-600 cm^{-1} range, we observe a downshift of A_{1g} mode for different chirality

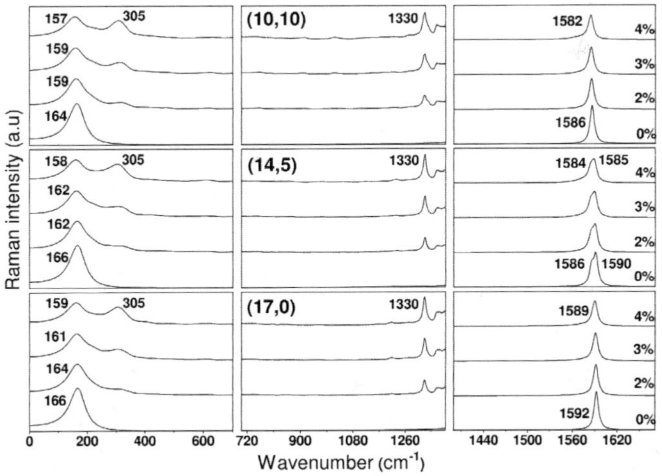

Figure 5. ZZ-polarized Raman spectra of different SWNT chirality as a function of hexa-vacancies defects concentration.

of SWNT, and a new mode appears near 305 cm^{-1} and keeps the same value independently of the SWNT chirality. The eigenvector displacements of the modes located around 305 cm^{-1} and 1330 cm^{-1} in (10,10) SWNT with hexa-vacancy defect, are given in Figure (3-B). These vectors allow us to identify the radial displacements of atoms belonging to the hexadeca-membered ring: all atoms in the hexa-vacancy move in phase in the radial direction and the mode is fully symmetric. This mode could be used to estimate hexa-vacancy defect of nanotubes.

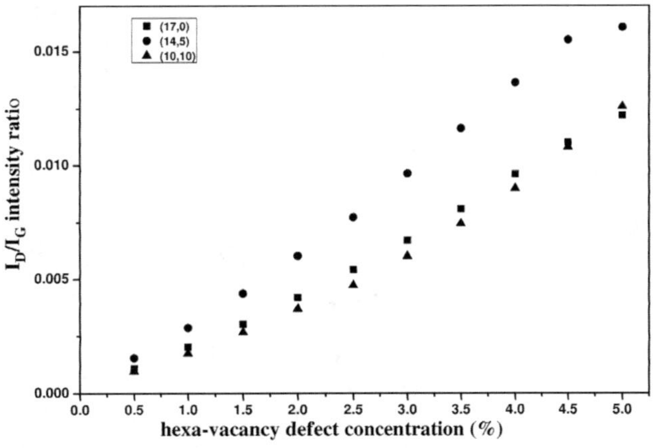

Figure 6. The evolution of the intensity ratio I_D/I_G as a function of the hexa-vacancies defects concentration in armchair (triangle), zigzag (square), and chiral (circle) nanotubes.

Raman spectroscopy is a very effective method to characterize the detailed bonding structure of carbon nanomaterials. The intensity ratio of the tangential mode of SWNTs (G-band) to the disorder-induced D-band centered at around 1330 cm^{-1} (I_D/I_G) provides a parameter that can be used for quantifying disorder in SWNT. The G band is a result of in-plane vibrations of sp^2 bonded carbon atoms whereas the D band is due to out of plane vibrations attributed to the presence of structural defects. Figure 6 shows a plot of (I_D/I_G) versus hexa-vacancy defect concentration for different chirality of nanotube (armchair (10,10), zigzag (17,0) and chiral (14,5)).

4. Conclusion
The nonresonant Raman spectra of SWNT with penta- and hexa-vacancies defects have been calculated using bond-polarizability model combined with the spectral moments method. For the hexa-vacancy defect in SWNT, we found that (i) The RBMs and G-mode of SWNTs are strongly dependent on the concentration of defects on the wall of nanotube. (ii) Two new modes are appeared due to hexa-vacancy defect on the D-band. The I_D/I_G ratio can be used to investigate the correlations between the Raman spectra and the defect structure of SWNT. We found that the low-energy Raman profiles showed more variations with the introduction of the defects in the wall of nanotube, whereas the high-energy profiles did not change appreciably depending on the defect concentration. The result of Raman spectra didn't significantly depend on the tube chirality. Our work provides benchmark theoretical results to understand the experimental Raman spectra of SWNTs with defect.

Acknowledgments
The works was supported by a CNRST-Morocco agreement.

References
[1] S. Iijima 1991 Helical microtubules of graphitic carbon, Nature, 354, 56-58.
[2] Wilder, Jeroen WG and Venema, Liesbeth C and Rinzler, Andrew G and Smalley, Richard E and Dekker, Cees 1998 Electronic structure of atomically resolved carbon nanotubes, Nature, 391, 59-62.
[3] Ruoff, Rodney S and Qian, Dong and Liu, Wing Kam 2003 Mechanical properties of carbon nanotubes: theoretical predictions and experimental measurements, Elsevier, 4, 993-1008.
[4] Mintmire. JW and White. CT 1995 Electronic and structural properties of carbon nanotubes, Carbon, 33, 893-902.
[5] Joselevich, Ernesto 2004 Electronic Structure and Chemical Reactivity of Carbon Nanotubes: A Chemist's View, Chem.Phys.Chem, 5, 619-624.
[6] Zhang, Daihua and Ryu, Koungmin and Liu, Xiaolei and Polikarpov, Evgueni and Ly, James and Tompson, Mark E and Zhou, Chongwu 2006 Transparent, conductive, and flexible carbon nanotube films and their application in organic light-emitting diodes, NANO LETTERS, 9, 1880-1886.
[7] Das, Rasel and Ali, Md Eaqub and Hamid, Sharifah Bee Abd and Ramakrishna, Seeram and Chowdhury, Zaira Zaman 2014 Carbon nanotube membranes for water purification: a bright future in water desalination, 336, 97-109.
[8] Ng, Kai-Wern and Lam, Wei-Haur and Pichiah, Saravanan 2013 A review on potential applications of carbon nanotubes in marine current turbines, Renewable and Sustainable Energy Reviews, 28, 331-339.
[9] Bandaru, R. Prabhakar 2007 Electrical properties and applications of carbon nanotube structures, Journal of nanoscience and nanotechnology, 7, 1239-1267.
[10] Rahmani, A and Sauvajol, J-L and Cambedouzou, J and Benoit, C 2005 Raman-active modes in finite and infinite double-walled carbon nanotubes, PHYSICAL REVIEW B, 71, 125402.
[11] Rahmani, A and Sauvajol, J-L and Rols, S and Benoit, C 2002 Nonresonant Raman spectrum in infinite and finite single-wall carbon nanotubes, PHYSICAL REVIEW B, 66, 125404.
[12] Umeno, Yoshitaka and Kitamura, Takayuki and Kushima, Akihiro 2004 Theoretical analysis on electronic properties of zigzag-type single-walled carbon nanotubes under radial deformation, Computational Materials Science, 30, 283-287.
[13] Hone, J and Llaguno, MC and Biercuk, MJ and Johnson, AT and Batlogg, B and Benes, Z and Fischer, JE 2002 Thermal properties of carbon nanotubes and nanotube-based materials, Applied Physics A, 74, 339-343.
[14] Ding, Feng and Bolton, Kim and Rosen, Arne 2004 Nucleation and growth of single-walled carbon nanotubes: A molecular dynamics study, The Journal of Physical Chemistry B, 108, 17369-17377.
[15] Yuan, Qinghong and Xu, Zhiping and Yakobson, Boris I and Ding, Feng 2012 Efficient defect healing in catalytic carbon nanotube growth, PHYSICAL REVIEW LETTERS, 108, 245505.
[16] Mawhinney, Douglas B and Naumenko, Viktor and Kuznetsova, Anya and Yates Jr, John T and Liu, J

and Smalley, RE 2000 Surface defect site density on single walled carbon nanotubes by titration, Chemical Physics Letters, 324, 213-216.

[17] Choi, Hyoung Joon and Ihm, Jisoon and Louie, Steven G and Cohen, Marvin L 2000 Defects, quasibound states, and quantum conductance in metallic carbon nanotubes, Physical Review Letters, 84, 2917.

[18] Dementev, Nikolay and Osswald, Sebastian and Gogotsi, Yury and Borguet, Eric 2009 Purification of carbon nanotubes by dynamic oxidation in air, Journal of Materials Chemistry, 19, 7904-7908.

[19] Moliver, SS and Zimagullov, RR and Semenov, AL 2011 FractureInduced Stone-Wales Defect Generation in Carbon Nanotube, Technical Physics Letters, 37, 678.

[20] Ma, Jie and Alfe, Dario and Michaelides, Angelos and Wang, Enge 2009 Stone-Wales defects in graphene and other planar sp2-bonded materials, PHYSICAL REVIEW B, 80, 033407.

[21] Robertson, Alex W and Montanari, Barbara and He, Kuang and Allen, Christopher S and Wu, Yimin A and Harrison, Nicholas M and Kirkland, Angus I and Warner, Jamie H 2013 Structural reconstruction of the graphene monovacancy, ACS Nano, 7 (5), 4495-4502.

[22] He, HY and Pan, BC 2008 Characteristic vibrational modes of a single vacancy in a zigzag carbon nanotube, PHYSICAL REVIEW B, 77, 073410.

[23] Brcic, Marino and Canadija, Marko and Brnic, Josip 2015 Influence of waviness and vacancy defects on carbon nanotubes properties, Procedia Engineering, 100, 213-219.

[24] Saidi, Wissam A and Norman, Patrick 2014 Spectroscopic signatures of topological and diatom-vacancy defects in single-walled carbon nanotubes, Physical Chemistry Chemical Physics, 16, 1479-1486.

[25] Berber, Savas and Oshiyama, Atsushi 2008 Atomic and electronic structure of divacancies in carbon nanotubes, PHYSICAL REVIEW B, 77, 165405.

[26] Muc, Aleksander and Banas, Aleksander and M. Chwal 2013 Free Vibrations of Carbon Nanotubes with Defects, Mechanics and Mechanical Engineering, 17, 157-166.

[27] P. Avouris 2007 Electronics with carbon nanotubes, Physics World, 20, 40-45.

[28] A. Jorio, M.S. Dresselhaus, and G. Dresselhaus 2008 Carbon Nanotubes: Advanced Topics in the Synthesis, Structure, Properties and Applications, Springer, ISBN: 0303-4216.

[29] C. Thomsen, S. Reich and J. Maultzsch 2004 Carbon Nanotubes: Basic Concepts and PhysicalProperties, Revista de Plasticos Modernos, ISBN: 3-527-40386-8.

[30] S. L. Mielke, D. Troya, S. Zhang, J.L. Li, S. Xiao, R. Car, R. S. Ruoff, G. C. Schatz, T. Belytschko 2004 The role of vacancy defects and holes in the fracture of carbon nanotubes, Chemical Physics Letters, 390, 413-420.

[31] K. Suenaga, H. Wakabayashi, M. Koshino, Y. Sato, K. Urita, and S. Ijima 2007 Imaging active topological defects in carbon nanotubes, Nature Nanotechnology, 2, 358.

[32] A. Hashimoto, K. Suenaga, A. Gloter, K. Urita1, and S. Iijima 2004 Direct evidence for atomic defects in graphene layers, Nature 430, 870.

[33] C. Gómez-Navarro, PJ. De Pablo, J. Gómez-Herrero, B. Biel, FJ. Garcia-Vidal, A. Rubio and F. Flores 2005 Tuning the conductance of single-walled carbon nanotubes by ion irradiation in the Anderson localization regime, Nature materials, 4, 534.

[34] Blanca Biel, F. J. Garca-Vidal, Angel Rubio, and Fernando Flores 2005 Anderson localization in carbon nanotubes: defect density and temperature effects, Physical review letters, 95, 266801.

[35] Andrews R., Jacques D., Qian D., Dickey E.C 2001 Purification and structural annealing of multiwalled carbon nanotubes at graphitization temperatures, Carbon 39: 1681-1687.

[36] Mawhinney D.B., Naumenko V., Kuznetsova A., Yates Jr. J.T., Liu J., Smalley R.E. 2000 Surface defect site density on single walled carbon nanotubes by titration, Chemical Physics Letters 324: 213-216.

[37] Nardelli M.B., Fattebert J.L., Orlikowski D., Roland C., Zhao Q., Bernholc J. 2000 Mechanical properties, defects and electronic behavior of carbon nanotubes, Carbon 38: 1703-1711.

[38] Terrones M., Botello-Mendez A.R., Campos-Delgado J., Lopez-Urias F., Vega-Cantu Y.I., Rodriguez-Macias F.J. 2010 Graphene and graphite nanoribbons: Morphology, properties, synthesis, defects and applications, Nano Today 5: 351-372.

[39] Ladislav Kavan, Peter Rapta, Lothar Dunsch, Michael J. Bronikowski, Peter Willis, and Richard E. Smalley 2001 Electrochemical tuning of electronic structure of single-walled carbon nanotubes: in-situ Raman and Vis-NIR study, The Journal of Physical Chemistry B, 105, 10764-10771.

[40] M. J. Matthews, M. A. Pimenta, G. Dresselhaus, M. S. Dresselhaus, and M. Endo 1999 Origin of dispersive effects of the Raman D band in carbon materials, Physical Review B, 59, 6585-6588.

[41] Ch. Thomsen and S. Reich 2000 Double resonant Raman scattering in graphite, PHYSICAL REVIEW LETTERS, 85, 5214-5217.

[42] M. A. Pimenta, G. Dresselhaus, M. S. Dresselhaus, L. G. Cancado, A. Jorio and R. Saito 2007 Studying disorder in graphite-based systems by Raman spectroscopy, Physical chemistry chemical physics, 9, 1276-1290.

[43] Cançado, L Gustavo and Jorio, A and Ferreira, EH Martins and Stavale, F and Achete, CA and Capaz, RB and Moutinho, MVO and Lombardo, A and Kulmala, TS and Ferrari, AC 2011 Quantifying defects in graphene via Raman spectroscopy at different excitation energies, Physical chemistry chemical physics, 11, 3190–3196.

[44] Saito, R and Jorio, A and Souza Filho, AG and Dresselhaus, G and Dresselhaus, MS and Pimenta, MA 2001 Probing phonon dispersion relations of graphite by double resonance Raman scattering, PHYSICAL REVIEW LETTER. 88, 027401.

[45] F. Tuinstra and J. L. Koenig 2004 Raman spectrum of graphite, The Journal of Chemical Physics, 53, 1126-1130, (1970).

[46] L. Wirtz, A. Rubio 2004 The phonon dispersion of graphite revisited, Solid State Communications 131 (3-4) 141-152.

[47] J. M. Soler, E. Artacho, J. D. Gale, A. Garcia, J. Junquera, P. Ordejon, D. Sanchez-Portal 2002 The siesta method for ab initio order-n materials simulation, Journal of Physics: Condensed Matter 14 (11) 2745.

[48] Bell R J 1976 Methods in Computational Physics, Vol. 15, Academic,New York.

[49] S. Guha, J. Menendez, J. Page, G. Adams 1996 Empirical bond polarizability model for infullerenes, Physical Review B 53 (19) 13106.

[50] A. Shahabi, M. Ghassemi, S.M. M. Langroudi, H. R. Nejad, M. H. Hamedi 2010 Computational Materials Science 50, 586-594

IOP Conf. Series: Materials Science and Engineering **783** (2020) 012015 doi:10.1088/1757-899X/783/1/012015

Theoretical study of electronic properties of nitrogen doped carbon nanotubes

Fatimaezzahrae Allali[a], Hassane Chadli[a], and Abdelali Rahmani[a*]

[a]Laboratoire d'Etude des Matériaux Avancés et Applications (LEM2A), Université Moulay Ismaïl, FSM, BP 11201,Zitoune, Meknes,Morocco.

*Corresponding author : a.rahmani@fs-umi.ac.ma

Abstract. In this work, we have studied the effects of nitrogen (N) substitutional doping on the electronical properties of single wall carbon nanotube (SWCNT). The electronic density of states (eDOS) of these nanosystems has been calculated in the framework of tight-binding calculations method. The obtained results show that the semiconducting SWCNT could be a quasi-metallic one after N-substitution. We conclude that electronic properties of the N- doped nanotubes (SWCNNT) are sensitive not only to the concentration of nitrogen atoms but also to their distribution on the nanotube surface. Our main results are discussed in the light of the experimental data.

1. Introduction

The Ever since their discovery [1], carbon nanotubes (CNTs) have been proclaimed as the new wonder material of the future. Their extraordinary electronic and optical properties destine them to play an important role in nanoelectronics. Calculations show that the intrinsic electronic properties of SWCNTs can be either semiconducting or metallic, depending on their geometrical structures defined by the chirality and the radius of the tube [2]. The skill to generate defects or/and to introduce impurities into CNT could be the most promising way to change and modulate its electronic properties and thus have a significant bearing on a broad range of applications [3,4]. Recently, different works have been done addressing the changes of CNT properties induced by doping with hetero-atom.
Nitrogen substitution of carbon atoms in NTCs modified the electrical and structural properties of these nanomaterials [5, 19] [6–8]. Both experimental and theoretical studies stated that the hetero-atom doping has an important impact upon the electronic structure of the CNTs [9,10].

The properties of SWCNTs are very sensitive to the tube diameter and chirality, which are depending on the synthesis conditions. A modification of these properties by controllably placing hetero-atoms on the tube surface leads to huge technological applications, which justifies the number of experimental and theoretical investigations focused on this topic. In fact, the electronic properties of SWCNTs can be modified by manipulations the Fermi level, which can be done by chemical doping. It was found that the introduction of hetero-atoms such as boron or nitrogen into SWCNTs offers the possibility of tailoring their structural and electronic properties [11–17]. Recently, SWCNTs doped with nitrogen or boron have been synthesized by laser-ablation [18], arc-discharge [19] and substitution reaction methods [20]. The N doped carbon nanotubes are produced by in situ nitrogen substitution and are used, singularly or in multiples, in power transmission cables, in solar cells, in batteries and in composites [21–25] [26].

The electronic properties of individual nitrogen doped SWCNTs (SWCNNTs) have been calculated by density functional theory [27] and tight-binding approach [28].
In this paper, we study the electronic properties of SWCNNTs, by using a force constants model in which the interatomic force constants up to the fourth-nearest-neighbor interactions are fitted to experimental data, to explore the basic properties of nitrogen doping. For this purpose, we perform calculations of the density of states (DOS) and dispersion curves.

2. Models and method

Density functional theory (DFT) methods are the standard and the most used theoretical techniques for electronic structure calculations. The formalism of the DFT and its extension to the chemical reactivity indexes are subject of intensive research and many empirical concepts [29–36]. The method is chosen for a huge range of applications. Density-functional based tight-binding (DFTB) is an approximate method based on the density functional framework which does not require large amounts of empirical parameters. These parameters are obtained from DFT calculations of a few molecules per pair of atom typesTight-Binding (TB) and Force Constant models are semi-empirical methods that do not require a fully self-consistent process for the computation of electron and phonon dispersion curves, and thus are much faster compared to ab initio calculations. However, they require the prior knowledge of some given empirical parameters which are called onsite and hopping energies in electron study and force constant parameters in phonon study. These parameters are usually adjusted so that dispersion results from TB (or Force Constant) simulation fit well with that of ab initio calculation or experiments. Great successes were evidenced from many works using TB methods on many materials and devices, including carbon nanomaterials [37–42]. In this study our approach is based on TB methods correlated to ab initio calculations to describe the electronic properties of nitrogen doped SWCNT and the used TB parameters are $t_{CC} = 2.85$eV and $t_{CN} = 2.96$eV [43]. We focus on the electronic properties of different achiral nitrogen doped carbon nanotubes of different diameters. The variation of the band gap energy with the concentration of N atoms is investigated.

3. Results and discussion

In this section, we report calculation results obtained for the eDOS and dispersion curves of armchair and zigzag single walled carbon nitride nanotubes (SWCNNTs) with different diameters and for different nitrogen concentrations.

3.1 Band structure and eDOS of nitrogen doped carbon nanotubes

In Fig.1, we plot the dispersion curves (on the left) of undoped (17,0) SWCNT which is a typical semiconducting carbon nanotube. As shown, the Fermi level is centered at zero energy with a bandgap energy between conduction and valence bands close to Eg=0.6eV These dispersion relations of a particular nanotube provide excellent information about the electronic properties of that nanotube, however, one may only require the eDOS which is deduced from the dispersion relations. Fig.1 (on the right) shows the electronic DOS of undoped (17,0) nanotube characterized by two separated large bands at low and at high energy.

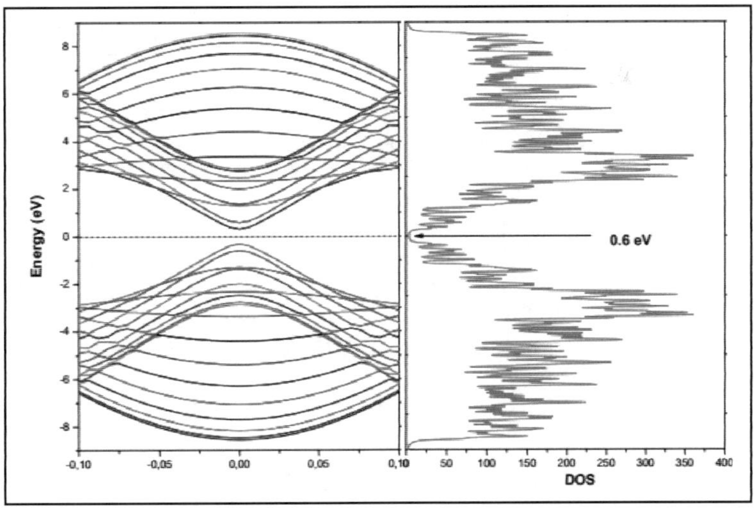

Fig. 1. Dispersion curves and density of states calculated for undoped (17,0) zigzag tube (τ =0 %).

When the concentration of nitrogen increases the profiles of the dispersion curves and eDOS are slightly affected, however one can see that the gap energy of the nanotube changes when C atoms are substituted by N ones on the surface of the tube. This behavior is well shown in Fig.2 where we plot the dispersion curves and eDOS calculated for doped SWCNNT at doping rate τ = 5%. At this concentration, the calculated bandgap energy is close to Eg =0.63eV. To analyze effects of nitrogen concentration (the ratio between the numbers of N atoms and C ones) on the electronic structure of SWCNT, we display in Fig.3, the dependence of the bandgap with the doping rate τ (in %) for (10,10) N-doped carbon nanotube. We note that we have performed calculations on samples with low nitrogen density limit where N atoms are considered at impurity level. Our results are comparable with the results found in the following references [44,45].

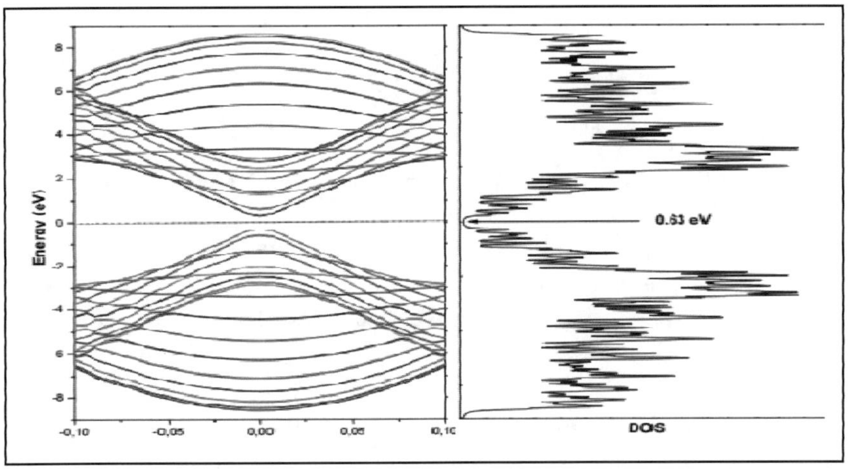

Fig. 2. Band structure and density of states calculated for (17,0) zigzag tube for τ = 5 %.

As we can see from Fig.3, the electrical conduction behavior of the nanotube changes from a metal to semiconductor for doping rate τ ranging from 0% with a gap energy E_g=0meV to 12% with E_g=24.2meV. In table 1, we give the calculated values of the bandgap for some doping rate and we state that for (10,10) SWCNNT this parameter increases linearly with rate doping .

Table 1. Gap energy variation with doping rate for (10,10) armchair SWCNNTs

Rate doping(%)	Gap(meV)
0	0
2	6,8
4	10,7
6	13,4
8	18,2
10	22,5
12	24,2

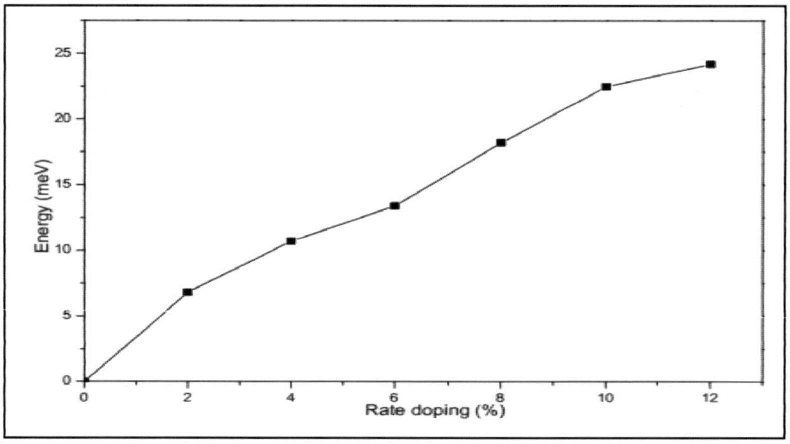

Fig. 3. Band gap in function the rate doping for (10,10) armchair tube. The Fermi level is located at zero energy.

3.2 The gap variation with diameter of nitrogen doped carbon nanotube

In Fig.4, we plot the calculated bandgap of SWCNNT versus diameter for (17,0) (panel a) and (10,10) (panel b) nanotubes with doping rates close to 8% and 5%, respectively. One can observe that for a given doping rate, the gap energy of armchair and zigzag SWCNNTs decreases when the diameter increases and can be represented by the relation: E_g=A/D with A is depending on the doping rate, together with A=0.14eV.nm and A=7.47eV.nm for armchair tube (a) and zigzag tube (b) respectively. This behavior is analogous to the dependence of the RBM frequency with diameter stated for SWCNT whatever the chirality and the conductivity of the nanotube. These results state that metallic armchair nanotube becomes semiconductor under nitrogen doping.

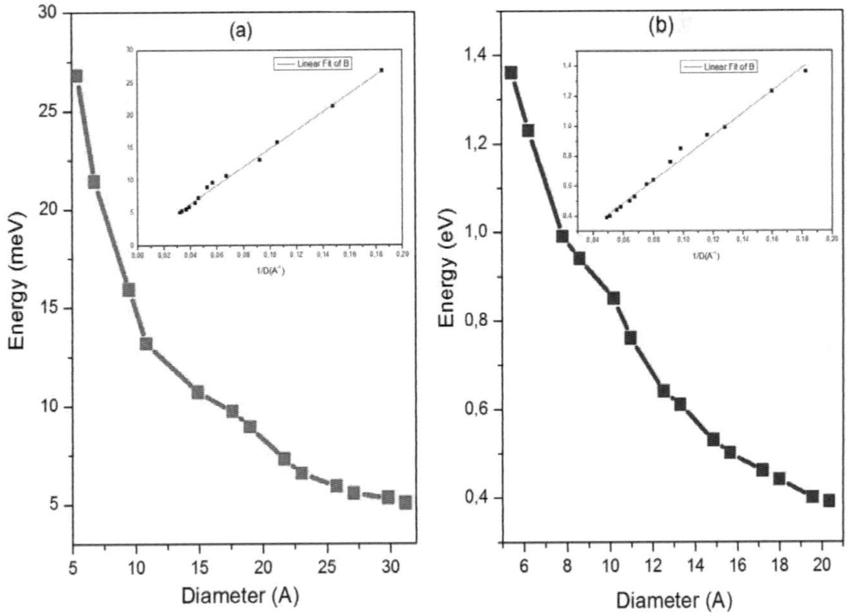

Fig. 4. The gap variation with diameter of nitrogen doped carbon nanotube for armchair tube at τ=8% (a) and zigzag tube τ=5 % (b). In inset : fit of this behavior.

4. Conclusion

In summary, we have performed band structure calculations on nitrogen doped carbon nanotubes within Tight- Binding approach. Due to nitrogen substitution doping effects, achiral nanotubes exhibit bandgap values linearly depending on the doping rate. The diameter effects study shows that the band gaps of both zigzag and armchair tubes decrease when the diameter increases. We also note that armchair nanotubes have zero gap but when the doping rate increases these nanomaterials switch from metal to semiconductor. These results may provide guidance in practical engineering applications.

References

[1] Iijima S (1991) Nature London **56** 354.
[2] Mintmire J.W, Dunlap B.I, White C.T (1992) Phys.Rev.Lett, **68** 631.
[3] Lowtner J.E (1998) Phys. Rev. B **57** 5724.
[4] Terrones M et al. (2002) Appl. Phys. A **74** 355.
[5] Sen R, Satishkumar B.C, Govindaraj A, Harikumar K.R, Raina Raina G, Zhang J, Cheetham A.K, Rao C.N.R (1998) Chem. Phys. Lett. **671** 287.
[6] Sekhar C. Ray, Mbiombi b W, Papakonstantinou c P (2014) Current Applied Physics **14.**
[7] Lim S. H, Elim H. I., Gao X. Y, Wee A. T. S, Ji W, Lee J. Y, and Lin (2006) J PRB **73,** 045402.
[8] Park, Hong-Lae and YI, Sung and Chung, Yong-Chae (2008) Japanese Journal of Applied Physics **47** 5062 .
[9] Suenaga K, Yudasaka M, Colliex C, Iijima S (2000) chem. Phys. Lett. **316 .**

[10] Kotakoski J, Pomoell A. V, Krasheninnikov A. V, Nordlund K (2005) Nucl. Instrum.Methods **228.**
[11] Stephan O, Ajayan PM, Colliex C, Redlich Ph, Lambert JM (1994) Science **266**.
[12] Carroll DL, Redlich Ph, Blase X, Charlier J-C, Curran S, Ajayan PM (1998) Phys Rev Lett **81** .
[13] Terrones M, Hsu WK, Schilder A, Terrones H, Grobert N, Hare JP (1998) Appl Phys A **66.**
[14] Kurt R, Karimi A (2001) Chem.phys.chem **6**.
[15] Hsu WK, Firth S, Redlich P, Terrones M, Terrones H, Zhu YQ (2000) J Mater Chem **6.**
[16] Golberg D, Bando Y, Kurashima K, Sato T (2001) Diamond Relat Mater **1** .
[17] Redlich P, Loeffler J, Ajayan PM, Bill J, Aldinger F, Ruhle M (1996) Chem Phys Lett **260.**
[18] Gai PL, Stephan O, McGuire K, Rao AM, Dresselhaus MS, Dresselhaus G (2004) J Mater Chem **4** .
[19] Glerup M, Steinmetz J, Samaille D, Stephan O, Enouz S, Loiseau A (2004) Chem Phys Lett **387** .
[20] Golberg D, Bando Y, Han W, Kurashima K, Sato T (1999) Chem Phys Lett **308** .
[21] Qu, G. X.; Cheng, J. L.; Li, X. D.; Yuan, D. M.; Chen, P. N.; Chen, X. L.; Wang, B.; Peng, H. S. A. (2016) Adv. Mater. **28.**
[22] Pan, S. W.; Ren, J.; Fang, X.; Peng, H. S. (2016) Adv. Energy Mater. **6.**
[23] Ni, J. F.; Wang, W. C.; Wu, C.; Liang, H. C.; Maier, J.; Yu, Y.; Li, L. (2017) Adv. Mater. **29.**
[24] Zhang, N.; Liu, Y. C.; Lu, Y. Y.; Han, X. P.; Cheng, F. Y.; Chen (2015) J. Nano Res. **8.**
[25] Zhao, F. P.; Gong, Q. F.; Traynor, B.; Zhang, D.; Li, J. J.; Ye, H. L.; Chen, F. J.; Han, N.; Wang, Y. Y.; Sun, X. H. et al. (2016) Nano Res. **9.**
[26] Wang, S. Q.; Xia, L.; Yu, L.; Zhang, L.; Wang, H. H.; Lou, X. W. D. (2016) Adv. Energy Mater. **6.**
[27] Takashi Koretsune and Susumu Saito (2008) Phys. Rev. B **77** .
[28] Yoshiyuki Miyamoto, Angel Rubio, Steven G. Louie, and Marvin L. Cohen (1994) PRB **50** 24.
[29] Argaman, N.; Makov, G.; Am. (2000) J. Phys. **68**, 69.
[30] Chermette, H.; Coord. (1998) Chem. Rev. **180**.
[31] Chermette, H. (1999) J. Comput. Chem. **20**.
[32] Kohn, W.; Becke, A. D.; Parr, R. G. (1996) J. Phys. Chem. **100.**
[33] Ladeira, A. C. Q.; Ciminelli, V. S. T.; Duarte, H. A.; Alves, M. C. M.; Ramos, A. Y. (2001) Geochim. Cosmochim. Acta **65.**
[34] De Proft, F.; Geerlings, P. (2001) Chem. Rev. **101.**
[35] Geerlings, P.; De Proft, F.; Langenaeker, W. (2003) Chem. Rev. **103.**
[36] Duarte, H. A. (2001) Quim. Nova **24.**
[37] Wallace P. R. (1947) Phys. Rev. **71.**
[38] Tran V.T., Saint-Martin J. and Dollfus P. (2014) Appl. Phys. Lett. **105.**
[39] Cresti A., Grosso G. and Parravicini G. P. (2008) Phys. Rev. B **77.**
[40] Rainis D, Taddei F, Dolcini F, Polini M and Fazio R (2009) Phys. Rev. B **79**.
[41] Grassi R, Poli S, Gnani E, Gnudi A, Reggiani S and Baccarani G (2009) Solid. State. Electron. **53.**
[42] Tran V-T, Saint-Martin J and Dollfus P (2015) Sci. Technol. **30**.
[43] Seol G and Guo (2011) J *Appl. Phys. Lett.* **98.**
[44] Charlier J.-C, Ebbesen T. W, and Lambin Ph (1996) Phys. Rev. B **53**.
[45] Blase X, Charlier J.-C, De Vita A, and Car R (1997) Appl. Phys. Lett. **70**.

IOP Conf. Series: Materials Science and Engineering **783** (2020) 012016 doi:10.1088/1757-899X/783/1/012016

Non resonant Raman spectra of homogeneous and inhomogeneous individual dimers of Single-walled Carbon Nanotubes (SWNTs)

M. Boutahir, S.A. Ait Abdelkader, B. Fakrach, AH. Rahmani, H. Chadli, A. Rahmani

Laboratoire d'Etude des Matériaux Avancés et Applications (LEM2A), Université Moulay Ismail,FSM-ESTM, BP 11201, Zitoune, 50000 Meknes, Morocco

E-mail: mourad.boutahir@gmail.com

December 2019

Abstract. We study the effects of the coupling between single-walled carbon nanotubes (SWNTs) on the phonons in homogeneous (identical nanotubes) and inhomogeneous (different nanotubes) dimers of SWNTs. Bond polarization theory and the spectral moments method (SMM) are used to calculate the non-resonant Raman frequencies of the breathing-like modes (BLMs) and tangential-like ones (TLMs). The dependence of the phonons on different structural parameters such as diameter and chirality is investigated. We focus on the dependence of the so-called Breathing-Like Modes (BLMs) of dimers originating from the van der Waals coupling of the Radial Breathing Modes (RBMs) of each individual tube forming the dimer. On the other hand, the TLM of dimer is reported and discussed. Distinct behaviors of the TLM are evidenced. The calculated of spectra are compared with the few experimental Raman data obtained on individual dimers.

1. Introduction.

Many researchers have devoted their efforts to study carbon nanotubes [1] and graphene [2] due to their various remarkable chemical, physical and mechanical properties [3]. These characteristic properties allow great potential applications [1, 4] of carbon nanotubes (CNTs) in many fields, such as photocatalysis [5], medicine [6], nanoscale electronics [7], optoelectronic devices [8], hydrogen storage [9, 10], mechanical systems [11], electrochemical devices [12], composites [13], SEM probes [14] and field emission devices [15, 16]. Carbon nanotubes exist in a wide variety of forms, such as single-walled carbon nanotubes (SWNTs), double-walled carbon nanotubes (DWNTs) and dimer of SWNTs.

For long time, it has been theoretically demonstrated a one-to-one correspondence between the atomic structure of a SWNT, especially its diameter, and the frequency of a low-frequency Raman active mode, the so-called radial breathing mode (RBM).

Recently, it was stated theoretically, and confirmed experimentally, that the RBM frequency vs Diameter (d) relationship can be expressed as: $\omega_{RBM}=\frac{228}{d}\sqrt{1+Cd^2}$ with $C = 0$ for individual environmental-free SWNTs (isolated SWNT) [21], and C $\neq$ 0 for SWNTs in interaction with an environment such as SWNTs sitting on a substrate, surrounded by different surfactants, or in bundles (Ref. [22] and references therein). In SWNT bundles, the origin of the C term mainly results from competition between the intertube interactions leading to an upshift of the radial breathing mode (called radial breathing-like mode (RBLM) in SWNT bundle) of about 8-16 cm^{-1}, depending on the tube diameter, with respect to the RBM of individual isolated SWNTs [17, 18, 19]. In SWNT bundles, in addition to the RBLM, a second mode is predicted in the low-frequency breathing mode range [18, 19, 20]. It originates from a double degenerate mode (E mode) in individual SWNT. Both bundle modes are fully-symmetric and can have mixed vibration patterns [18]. The most small bundle: the dimer of SWNTs, is characterized by weak van der Waals interactions between the neighboring carbons of its two constituting SWNTs. Dimer of two identical (non-identical) SWNTs, the so-called homogeneous (inhomogeneous) dimers, have three electronic configurations, namely: SC-SC, SC-M and M-M, where SC (resp. M) stands for semiconducting (resp. metallic) SWNT.

Individual carbon nanostructures, such as double-walled carbon nanotubes (DWNTs) and dimers of SWNT provides unique model systems for the study of van der Waals coupling at the nano scale [24]. Raman spectroscopy is known as a powerful tool for studying the vibrational properties of such carbon nanostructure [25], (other references have to be added) especially the dependence of these properties as a function of the diameters [26].

In a previous paper, using the spectral moments method (SMM) [27, 28, 29, 30], and in the framework of the bond-polarization theory, the polarized Raman spectra of DWNTs have been calculated. The dependence of the Raman spectrum of DWNTs, with the diameter and chirality of each constituting SWNTs, was investigated.

In the present work, the Raman spectra of other van der Waals-coupled carbon nanostructures, the homogeneous and inhomogeneous dimer of SWNTs, are calculated. The dependence of the Raman active modes on the diameters, chirality and length of each constituting SWNT is investigated. The results of calculations concerning the low-frequency breathing like modes (BLMs) of dimers of SWNTs are compared with the few available experimental Raman data.

IOP Conf. Series: Materials Science and Engineering **783** (2020) 012016 doi:10.1088/1757-899X/783/1/012016

2. Models and computational method.

The force constant model FCM model we employ involves a fourth nearest neighbor approximation. The force constant tensor describes the coupling between the ith and the jth carbon atom, which are N^{th} nearest-neighbor of each other[31].
(has to be precised) A Lennard-Jones potential $U_{LJ}(r) = 4\varepsilon\left[\left(\frac{\sigma}{r}\right)^{12} - \left(\frac{\sigma}{r}\right)^{6}\right]$ is used to describe the van der Waals interactions between the two constituting SWNTs of the dimers, with parameters $\epsilon = 2.964$ meV and $\sigma = 0.3407$ nm .

We calculate the Raman spectra of dimers of SWNT using the spectral moment's method (SMM). For SWNT, the SMM method is described in detail in Ref[29]. And the same approach is used in this work. Raman tensor of each vibrational mode is calculated in the framework of non-resonant bond-polarizability model[32]. The frequencies of Raman-active modes are derived from the position of the peaks in the calculated Raman spectra the intensity of the Raman spectrum is normalized with respect to the number of carbon in the sample under consideration. In our calculations, the common axis of the two tubes is along the Z axis.

3. Results.

In this section, the calculated Raman spectra of infinite homogeneous and inhomogeneous dimers of SWNTs are reported. The dependence of the Raman spectrum as function of diameter and chirality of the constituting SWNTs is investigated.

3.1. Infinite homogeneous Dimer of SWNTs.

We now turn to the study of For each investigated dimer, the Lennard-Jones intertube energy was minimized. The optimal intertube gap is found around 3.4Å.

In a first time, we focus on the Raman spectra of armchair-armchair dimers, identified as (n,n)-(n,n). In order to illustrate our approach, the breathing-like modes (BLM) (figure1-(A)) of the Raman spectra of (5,5)-(5,5), (10,10)-(10,10), (15,15)-(15,15) and (20,20)-(20,20) dimers are discussed.

The Raman spectrum of the (15,15)-(15,15) dimer displays two intense peaks located at 110 (BLM2) and 115 cm^{-1} (BLM3), and a weak peak at 82 cm^{-1} (BLM1). All these modes result from the coupling of all the breathing modes of isolated tubes. For infinite dimers, the eigenvector of each mode modes is easily obtained from the direct diagonalization of the dynamical matrix. By this way, we found that BLM2 mainly corresponds to the coupling of the radial breathing mode (RBM) of isolated (15,15) tube. The frequency of the RBM in isolated (15,15) armchair tube is calculated at $108cm^{-1}$,

IOP Conf. Series: Materials Science and Engineering **783** (2020) 012016 doi:10.1088/1757-899X/783/1/012016

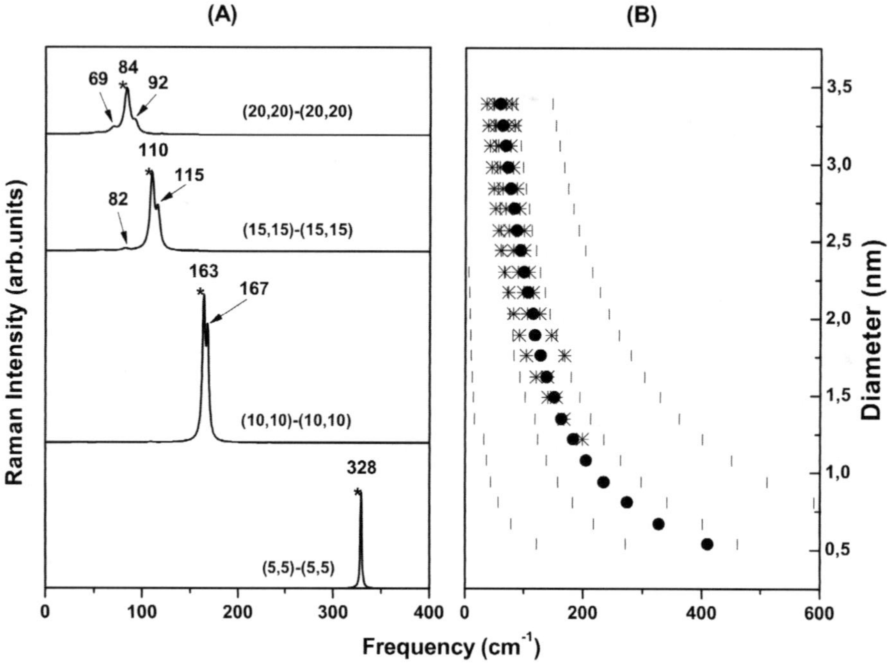

Figure 1. (A) The ZZ calculated Raman spectra of infinite (5,5)-(5,5), (10,10)-(10,10), (15,15)-(15,15) and (20,20)-(20,20) dimers in the BLM range. (B) Diameter dependence of the Raman-active mode frequencies for armchair-armchair dimers.

and, as expected, the calculations show a slight upshift of the frequency of the BLM2 in (15,15)-(15,15) dimer. The BLM1 and BLM3 peaks are assigned to hybridation between radial breathing modes and the E_n modes of individual (15,15) SWNT [18]. Notice that hybridation between the breathing modes increases with increasing the diameters of the tubes.

In figure1-(B) we discuss the dependence with the diameter of the breathing-like modes, in low diameter (0.5-1.2nm), one RBLM mode is found (full circle). at diameter 1.35nm another RBLM mode appear resulting to the coupling mode of both tubes(open star) due to the van der Waals interactions between the tubes. these interactions become more intersting when the diameter increases. Notice that as in SWNT, the BLM's are not significantly dependent on the chirality of the tube, that's the reason why we calculate only for armchair-armchair dimer(smaller unit cell than for chiral and therefore faster calculations).

We examine now the TLM range starting by Armchair-Armchair dimers. the corresponding spectra are shown in the figure2-(A) for different systems. As clear from figure2-(A), we observe here one peak of the tangential modes (TM's) of both systems. From calculations the Raman spectrum of (10,10)-(10,10) dimer displays TLM's at $1586cm^{-1}$. When the tube diameter increases, an upshift of TLM's modes

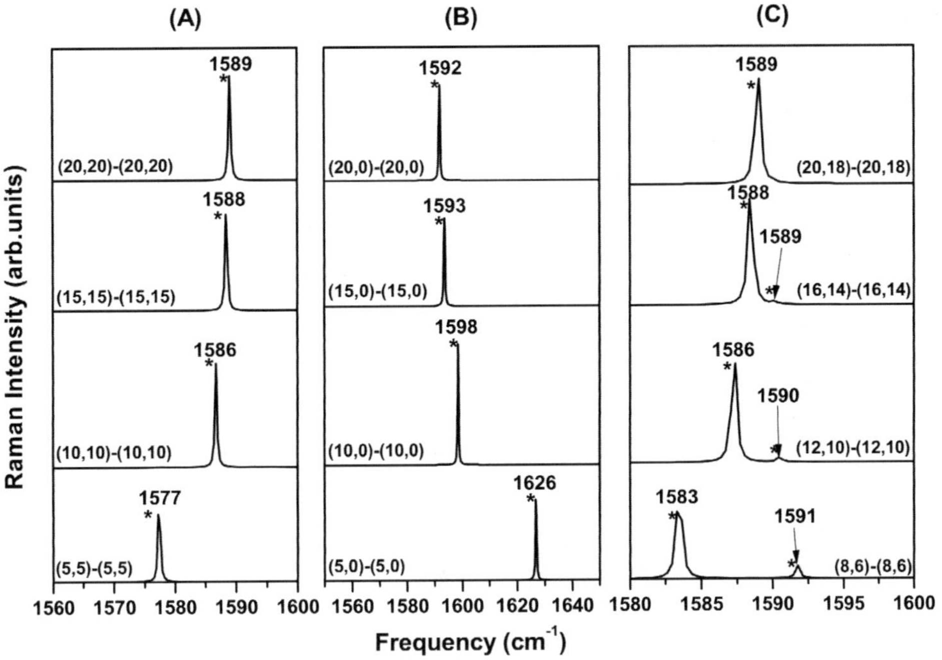

Figure 2. The ZZ calculated Raman spectra of infinite armchair-armchair (A), zigzag-zigzag (B) and chiral-chiral (C) dimers in the TLM ranges.

around $1589cm^{-1}$ of large diameter. In the Zigzag-Zigzag dimers the corresponding spectra are shown in the figure2-(B). Also the Raman spectra displays one peak of the tangential modes (TLM's) of both systems. The TLM frequency of (10,0)-(10,0) dimer is $1598cm^{-1}$. But on the contrary of Armchair-Armchair dimer when the tube diameter increases, a downshift of TLM's modes around $1592cm^{-1}$ of large diameter. On the other hand the Raman spectrum of a chiral-chiral systems displays two bands of the tangential modes (TLM's)(figure2-(C)). We deduced as in SWNT, the Raman spectrum in the TLM region is very sensitive to the chirality of the tubes.

In contrary to the breathing like modes range, the coupling modes don't exist in the TM region. For example, TLM of (10,10)@(10,10) is centered at 1586 cm^{-1}. In isolated (10,10) SWNT, Raman line is calculated at 1585 cm^{-1}. Also in a chiral-chiral systems like (12,10)-(12,10) one, the Raman spectrum presents two peaks (1586 and 1590cm^{-1}). In isolated (12,10) SWNT, Raman lines are calculated at (1585 and 1591 cm^{-1}). The comparison between the calculated Raman spectra shows that the Dimer spectra are a superposition of the two Raman spectra of the corresponding isolated tubes in this range.

IOP Conf. Series: Materials Science and Engineering **783** (2020) 012016 doi:10.1088/1757-899X/783/1/012016

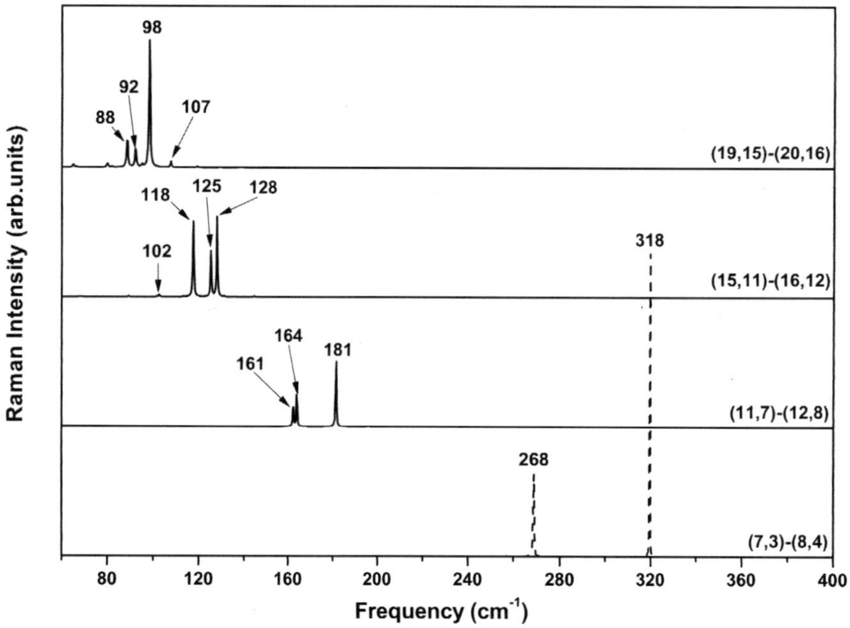

Figure 3. The dependance of diameter of ZZ calculated Raman spectra of dimers in the BLM range.

3.2. Infinite inhomogeneous Dimer of SWNTs.

We report the calculated Raman spectrum for infinite inhomogeneous dimers of SWNTs, the calculations are performed on infinite systems, we applied periodic boundary conditions on unit cells of the considred nanotubes forming the system.

In a first time, we focus on the Raman spectra of chiral-chiral dimers as function of diameter. In order to illustrate our approach, the breathing-like modes (BLM) (figure3) of the Raman spectra of (7,3)-(8,4), (11,7)-(12,8), (15,11)-(16,12) and (19,15)-(20,16) dimers are discussed.

The Raman spectrum of the (19,15)-(20,16) dimer displays four peaks located at 88 (BLM1), 92 cm^{-1} (BLM2), 98 cm^{-1} (BLM3) and 107 cm^{-1} (BLM4). All these modes result from the coupling of all the breathing modes of isolated tubes. For infinite dimers, we use the same method (direct diagonalization) to obtain the eigenvector of each mode. By this way, we found that BLM2 (resp.BLM3) mainly corresponds to the coupling of the radial breathing mode (RBM) of isolated (20,16) (resp.(19,15)) tube. The BLM1 and BLM4 peaks are assigned to hybridation between radial breathing modes and the E_n modes of individual SWNTs. Notice that hybridation between the breathing modes increases with increasing the diameters of the tubes as shown in figure3, in the small diameter ((7,3)-(8,4)) the Raman spectrum displays two peaks lacated at 268cm^{-1} and

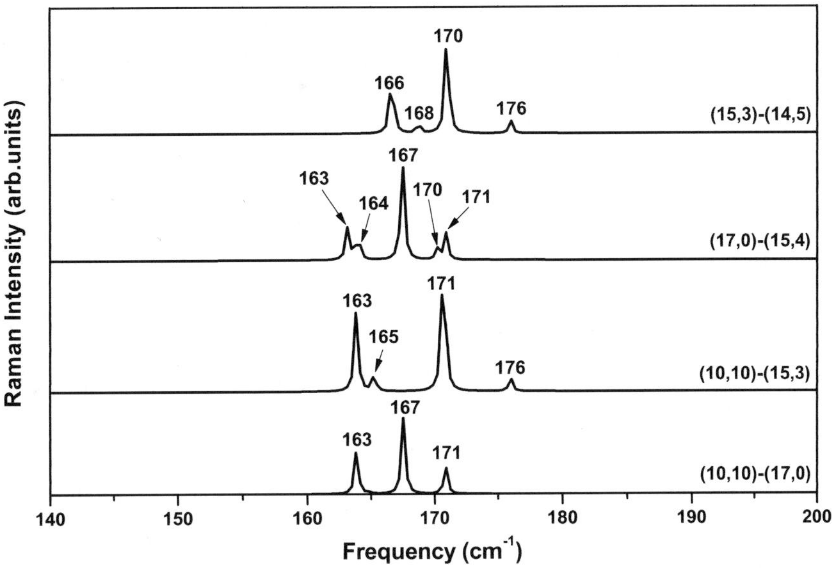

Figure 4. The ZZ calculated Raman spectra of different chirality of dimers in the BLM range.

$318cm^{-1}$, these modes corespond to the coupling of the RBM of isolated (8,4) and (7,3) tubes. The hybridation modes are not appear. When the diameter increases the hybridation modes increase.

In figure4 we present the ZZ polarized Raman spectrum of four configurations of dimers. Zigzag-Chiral dimer ((17,0)-(15,4)), Armchair-Zigzag dimer ((10,10)-(17,0)) , Armchair-Chiral dimer ((10,10)-(15,3)) and Chiral-Chiral dimer ((15,3)-(14,5)) in the BLM range. All diameters of the tubes close to 1.35nm. The dependence of chirality is discussed.

In Armchair-Zigzag dimer ((10,10)-(17,0)), the Raman spectrum shows three peaks located at 163(BLM1), 167(BLM2) and $170cm^{-1}$(BLM3), the lower-frequency pair of peaks ($163cm^{-1}$ and $167cm^{-1}$) is assigned to vibrations of the (10,10) tube and the higher-frequency ($170cm^{-1}$) to the vibrations of (17,0) tube. To compare with the (10,10)-(10,10) dimer we found that the two peaks corresponding to (10,10)-(10,10) dimer is found in (10,10)-(17,0) dimer. In the (10,10)-(10,10) the $163cm^{-1}$ peak corresponding to the chybridation mode and $167cm^{-1}$ to the radial breathing mode of (10,10) tube. An upshift of radial breathing mode of (10,10) tube in (10,10)-(17,0) dimer by $5cm^{-1}$ and an upshift by only $1cm^{-1}$ in (10,10)-(10,10) dimer. Now in the (10,10)-(15,3) the Raman spectrum displays four peaks located at 163(BLM1),

165(BLM2), 170(BLM3) and 176cm^{-1}(BLM4), the lower-frequencies pair of peaks (163cm^{-1} and 165cm^{-1}) is assigned to vibrations of the (10,10) tube and the higher-frequency (170cm^{-1} and 171cm^{-1}) to the vibrations of (15,3) tube.

The BLM1 and BLM4 are a hybridation modes of both tubes, BLM2 corresponding to a radial breathing mode to (10,10) and BLM3 is a radial breathing mode to (15,3). We found a systematic upshift of the frequency of the BLM2 with respect to the frequency of the breathing mode in (10,10) SWNT by 3cm^{-1}. This upshift of the radial-breathing mode of (10,10) tube in dimers is controlled by the chirality of the other tube that forms the dimer with (10,10) tube.

In the TLM region, the Raman spectrum of different dimers presented in figure5 show four peaks are called TLM1, TLM2, TLM3 and TLM4. The first two frequencies correspond to the Transverse-Optical modes (TO), the second two modes correspond to Longitudinal-Optical modes (LO) modes of the semiconducting carbon nanotubes. For instance, the (7,3)-(8,4) dimer displays four bands located at 1576, 1580, 1594 and 1596 cm^{-1}, the two pair of peaks (1576 and 1594cm^{-1}) correspond to the vibrations of (7,3) tube and (1580 and 1596cm^{-1}) those of (8,4) tube. While the frequencies of the tangential modes in isolated (7,3) and (8,4) tubes are respectively calculated at (1584 and 1595 cm^{-1}) and (1586 and 1593 cm^{-1}). We notice that as in homogeneous dimers the calculated Raman spectra shows that the Dimer spectra are a superposition of the two Raman spectra of the corresponding isolated tubes in this range. In addition, our calculations state a changes of the tangential-like modes in Dimer, caused by intertube van der Waals interactions, with respect to SWNTs.

In figure6 the Raman spectrum of four systems are presented, the diameter of the tubes close to 1.35nm. We found two modes in armchair-zigzag ((17,0)-(10,10)) dimer, one mode of each tube 1586cm^{-1}) for (17,0) and 1592cm^{-1}) for (10,10). Three modes in armchair-chiral ((10,10)-(15,3)) 1586cm^{-1}) for (10,10) tube and two pair peaks (1585 and 1592cm^{-1})) for (15,3) tube, also three bands in zigzag-chiral ((17,0)-(15,4)) 1592cm^{-1}) for (17,0) tube and two pair peaks (1586 and 1593cm^{-1})) for (15,4) tube. In the chiral-chiral ((15,3)-(14,5)) dimer, four modes are found two pair peaks (1576 and 1592cm^{-1})) for (15,3) tube and two pair peaks (1583 and 1591cm^{-1})) for (15,4) tube. We notice that the TLM region is very sensitive to the chirality of the tubes in the TLM range as we found in homogeneous systems.

4. Discussions.

The (15,11)-(16,12) is an inhomogeneous dimer and the distance between them is close to d=3.4Å. Their chiral angles are boths close to 25°. In figure7 the Raman spectra displays the four peaks corresponding to A_{1g} modes for (15-11)-(16-12) dimer, we found four bounds (called BLM) located at 102cm^{-1} (BLM1), 118cm^{-1} (BLM2), 125cm^{-1} (BLM3) and 128cm^{-1} (BLM4).

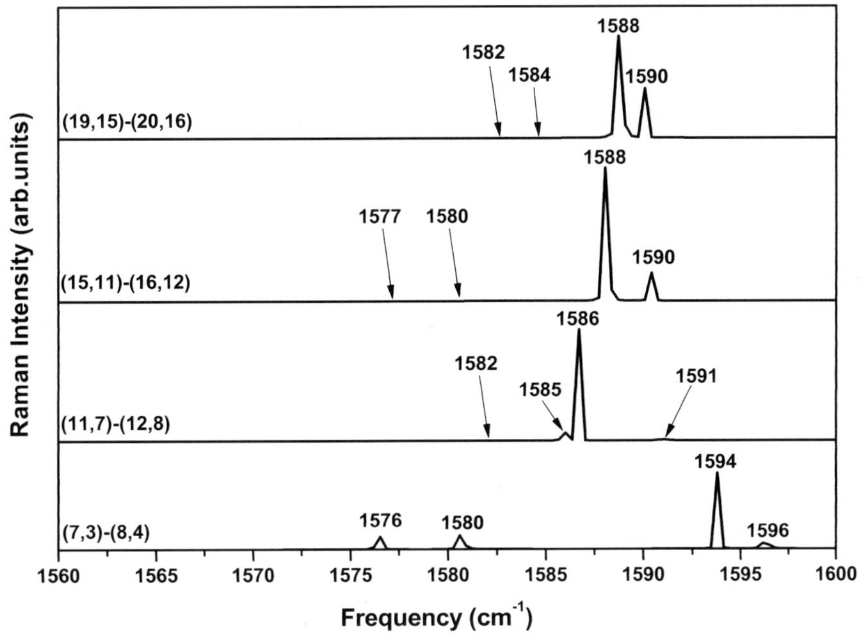

Figure 5. The dependance of diameter of ZZ calculated Raman spectra of dimers in the TLM range.

The presence of additional modes as compared to the two expected radial-breathing modes (RBM) is attributed to the lowering of the symmetry in dimers [34, 35]. It is tempting to assign the lower-frequency pair of peaks ($102cm^{-1}$ and $118cm^{-1}$) to vibrations of the (16,12) tube and the higher-frequency pair ($125cm^{-1}$ and $128cm^{-1}$) to the vibrations of (15,11) tube.

To analyze the experimental data in the light of the calculated features of our work, we consider some given experimental frequency from reference [36]. Recent Raman measurements on (15,11)-(16,12) Dimer, obtained by direct synthesis method, performed by [36], showed that RBM Raman spectrum is dominated by four peaks around 124, 131, 138, $146cm^{-1}$.

Débarre et al [37] have reported the Raman spectrum of (8,6)-(18,4) dimer with two peaks at $150cm^{-1}$ (BLM2) and $245cm^{-1}$ (BLM3). From our calculations the BLM2 and BLM3 are found around 140 and $235cm^{-1}$. We notice that in the last system ((8,6)-(18,4)) the BLM1 and BLM4 (hybridation modes) are very weak because the diameter of the tubes are small (0.95nm for (8,6) and 1.59nm for (18,4)).

We examine now the TLM range of Raman spectrum of the (15,11)-(16,12) dimer. The corresponding spectra is shown in the figure7. The TM frequencies are 1577(TLM1), 1580(TLM2), 1588(TLM3) and 1590 cm^{-1}(TLM4). The first two frequencies correspond

IOP Conf. Series: Materials Science and Engineering **783** (2020) 012016 doi:10.1088/1757-899X/783/1/012016

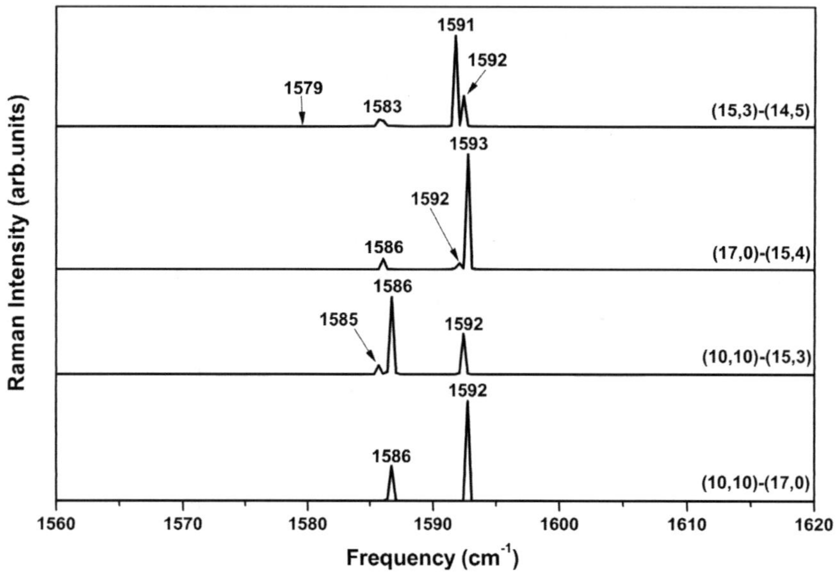

Figure 6. The ZZ calculated Raman spectra of different chirality of dimers in the TLM range.

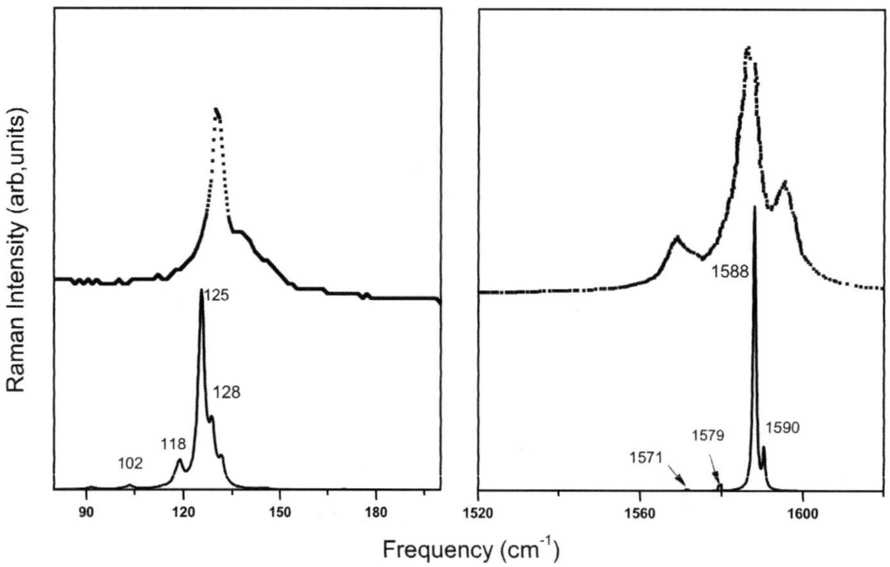

Figure 7. (top) experimental Raman spectrum (bottom) ZZ polarized Raman spectrum of (15,11)-(16,12) dimer.

to the TO modes, the second two to LO modes of the constituent semiconducting carbon nanotubes. As was shown in the figure2-(C), the TO mode's frequency show a strong dependence on the nanotube's diameter, while LO modes do not have a big change. In this system we assign the 1577, 1588cm^{-1} to a (15,11) nanotube and 1580, 1590cm^{-1} to a (16,12).

The frequencies of TO modes predicted for isolated individual SWNTs (15,11) and (16,12) are 1588cm^{-1} and 1589cm^{-1} respectively. These frequencies are upshifted with respect to the corresponding TO modes in the bundle (1577cm^{-1} and 1580cm^{-1}). We observe also that the difference between the frequency of the TO mode in (16,12) and (15,11) individual SWNTs is around 1cm^{-1}, that is slightly smaller than the difference calculated in the dimer (around 3cm^{-1}). In this assumption, for TO modes, the interaction between the tubes leads a downshift of frequency of the both TO modes and to an increase in their frequency separation. In this framework these two features could be considered as the signature of the coupling bewtween the TO modes of SWNTs in dimer.

Finally, a careful examination of the LO modes lead to a a weak evidence of the coupling between SWNTs in this dimer. Indeed, there is a small deviation of the LO frequencies of the dimer from those found in isolated individual tubes. assuming the frequencies of the LO mode in isolated individual SWNTs (15,11) and (16,12) as (1590cm^{-1} and 1591cm^{-1} respectively). These frequencies are downshifted by 1cm^{-1} with respect to the corresponding LO modes in the bundle (1588cm^{-1} and 1590cm^{-1}).

In table 1 we compare our calculations with the experimental BLM and TLM measured in Ref[38]. This values gives the best agreement between experimental and our calculations.

Table 1. comparison of BLM and TLM frequencies obtained with an expermental data from Ref[38].

	our works (cm^{-1})	ref[38] (cm^{-1})
BLM1	102	124
BLM2	118	131
BLM3	125	138
BLM4	128	146
TLM1	1577	1568
TLM2	1580	1571
TLM3	1588	1586
TLM4	1590	1596

5. conclusion

In this article, we investigated the non-resonant Raman study of Van der Waals coupled carbon nanostructures, namely, a bundle of two identical and non-identical SWNTs (Dimer). We showed that Van-der-waals coupling between the tubes lead to collective radial breathing-like modes (RBLM). The dependencies of these RBLMs as a function of diameter and chirality were discussed. Also we discussed the TLM region of dimer of two SWNTs we found two distinct behaviors of TLM: i) A clear shift of TLM frequencies of constituent tubes in comparison with those of SWNTs. ii) The spectrum of dimer was just a simple sum of TM of corresponding SWNTs. In order to compare with the experimental data we calculated the Raman spectrum of (15,11)-(16,12) dimer. We found that our calculations are in good agreement with experimental data.

6. ACKNOWLEDGMENT

The work was supported by the CNRS-France/CNRST-Morocco agreement.

7. references

[1] Lijima S *Helical microtubules of graphitic carbon Nature* 1991 **354**,56-58.
[2] Novoselov K S Geim A K Morozov SV JiangZhang Y D Dubonos S V Grigorieva I V Firsov A A (2004) *Science* **306** 666.
[3] Scheibe B Palen E Kalenczuk R (2010) *J Alloys Compd* **500** 117.
[4] Endo M Kim C Nishimura K Fujino T Miyashita K (2000) *Carbon* **38** 183.
[5] Chen C Liang Y Zhang W (2010) *J. Alloys Compd* **501** 168.
[6] Cheung W Pontoriero F Taratula O Chen A M He H (2010) *Adv. Drug Deliv. Rev* **62** 633.
[7] Hermann S Pahl B Ecke R Schulz S E Gessner T (2010) *Microelectronic. Eng* **87** 438.
[8] Yao Z Postma H W Balents L Dekker C (1999) *Nature* **402** 273.
[9] Rather S Naik M Hwang S Kim A Nahm K (2009) *J. Alloys Compd* **475** 17.
[10] Jain I P Jain P (2010) *J. Alloys Compd* **503** 303.
[11] Kuo-Jung Lee, Mao-Hsiang Hsu, Huy-Zu Cheng, Jason Shian-Ching Jang, San-Wo Lin, Cheng-Ching Lee, Sang-Chaur Lin, (2009) *J.Alloys Compd* **483** 389.
[12] Baughman R H Cui C Zakhidov A A Iqbal Z Barisci J N Spinks G M Wallace G G Mazzoldi A De Rossi D Rinzler A G Jaschinski O Roth S Kertesz M (1999) *Science* **284** 1340.
[13] Thostenson E T Ren A Chou T W (2001) *Compos. Sci. Technol* **61** 1899.
[14] de Jonge N (2009) *Adv. Imag. Electron. Phys* **156** 203.
[15] Darbari S Abdi Y Mohajerzadeh S Asl Soleimani E (2010) *Carbon* **48** 2493.
[16] Shim M Javey A Kam N W S Dai H J Am J (2001) *Chem. Soc* **123** 11512.
[17] Dresselhaus M S Eklund P C (2000) *Adv. Phys* **49** p.705
[18] Henrard L Popov V N Rubio A (2001) *Phys. Rev. B* **64** p.205403
[19] Sbai K Rahmani A Chadli H Sauvajol J- L *J. Phys.: Condens. Matter* **21** (2009) p.045302
[20] Popov V N Henrard L (2001) *Phys. Rev. B* **63** p.233407
[21] Liu X K Wang W Wu M Xiao F Hong X Aloni S Bai X Wang E Wang F (2011) *Phys. Rev. B* **83** p.113404
[22] Araujo P T Maciel I O Pesce P B C Pimenta M A Doorn S K Qian H Hartschuh A Steiner M Grigorian L Hata K Jorio A (2008) *Phys. Rev. B* **77** p.241403.
[23] Blancon J -CPaillet MTran H NThan X TAbera Guebron SAyari ASan Miguel APhan N -MZahab A -A Sauvajol J -LDelFatti NVallée F (2013) *Nat.Comm* **4** 2542.
[24] Dresselhaus M S Dresselhaus G Saito R Jorio A (2005) *Phys. Rep* **409** 47.
[25] Kavan L Rapta P Dunsh L Bronikowski M J Willis P Smalley R E (2001) *J. Phys. Chem. B* **105** 10764
[26] Astakhova T Yu Vinogradov G A Menon M (2003) *Chem. Bull* Int. Ed. **52** 823.
[27] Rahmani A Sauvajol J -L Cambedouzou J Benoit C (2005) *Phys. Rev. B* **71** 125402.
[28] Rahmani A Jund P Benoit C Jullien R (2001) *J. Phys: Condens, Matter* **13** 5413.
[29] Rahmani A Sauvajol J -L Rols S Benoit C (2002) *Phys.Rev.B* **66** 125404.
[30] Benoit C Royer E Poussigue G (1992) *J.Phys. Condens.Matt* **4** 3125.
[31] Wirtz L Rubio A (2004) *Solid State Commun* **131** 141.
[32] Guha S Menendez J Page J B Adams G B(1996) *Phys. Rev. B* **53** 13 106.
[33] Lu J Wang S (2007) *Phys. Rev.B* **76** 233103.
[34] Sbai K Rahmani A Chadli H Sauvajol J-L (2009) *J.Phys: Condens.Matter* **21** 045302.
[35] Henrard L Popov V Rubio A (2001) *Phys. Rev. B* **64** 205403.
[36] Levshov D Michel T Paillet M Than X T Tran H N Arenal R Rahmani A Boutahir MZahab A -A Sauvajol J -L (2014, January) *In MRS Proceedings* **Vol. 1700** pp. mrss14-1700. Cambridge University Press.
[37] Débarre A Kobylko M Bonnot A M Richard A Popov V N Henrard L Kociak M (2008) *Physical review letters* **101(19)** 197403.
[38] Levshov D (2013) Raman modes in index-identified individual single-walled and multi-walled carbon nanotubes. Unpublished PhD thesis. University of Montpellier 2.

IOP Conf. Series: Materials Science and Engineering **783** (2020) 012017 doi:10.1088/1757-899X/783/1/012017

Wall thikness effects on the infrared spectra of multi-walled carbon nanotubes

Mourad Boutahir[1,*], Brahim Fakrach[1, 2], Abdelhai Rahmani[1], Khalid Sbai[1], Hassan Chadli[1], and Abdelali Rahmani[1]

[1]Laboratoire d'Etude des Matériaux Avancés et Applications, Université Moulay Ismail, Faculté des Sciences Meknes, Morocco
[2]Laboratoire de Physique Théorique et Appliquée, Université Sidi Mohammed Ben Abdellah, Faculté des Sciences Dhar El Mahraz Fez, Morocco
*Corresponding author e-mail: mourad.boutahir@gmail.com

Abstract. In this theoretical work, we investigate the infrared vibrational features of multi-walled carbon nanotubes (MCNTs). We calculate the polarized infrared spectra of multi-walled carbon nanotubes using the spectral moment's method. This original approach allows us to consider MCNTs with a large number of walls. We discuss the evolution of the low, intermediate and high wavenumber regions of these spectra as a function of the inner diameter and the size of MCNTs. Both the XX and ZZ polarisations are considered. Our results provide benchmark theoretical data for understanding the experimental infrared spectra of MCNTs.

1. Introduction

Since the time of their discovery [1,2], many efforts were taking to study Carbon nanotubes (CNTs). They became more and more popular to the scientists due to their various remarkable chemical, physical and mechanical properties [3,4]. These characteristics open the way for their use in many applications [3-5]. Initially, the electronic and optical properties give them the great interest. The ability to engineer them offers exciting possibilities for photovoltaic with hydrogen storage applications [6,7]. Many forms exist of CNTs, namely single-walled carbon nanotubes (SCNTs), double-walled carbon nanotubes (DCNTs) and multi-walled carbon nanotubes (MCNTs) [8].

The MCNTs forms like a concentric graphene layer rolled in a cylindrical form, the diameters are around tens of nanometers [9]. There are many developed synthesis methods to fabricate both SCNTs and MCNTs. The most promising method for producing CNTs is the catalytic vapor chemical deposition (CVD) [10]. Either SCNTs or MCNTs can be produced for many applications [11-15]. It depends on the operating conditions and the catalyst system used.

Raman spectroscopy has provided a powerful technique for studying the CNTs. It is used to investigate the vibrational properties of CNTs in relation to their structural and electronic properties [16,20]. Various kind of carbon nanotube forms can be analysed by the Raman spectroscopy including SCNTs, DCNTs and MCNTs. Since the Infrared activity is related to a dynamic dipole moment that is weak, the Infrared spectroscopy of CNTs has been neglected. Nevertheless, a surprising number of phonon mode bands are IR-active through transient dipole.

Various of optical spectroscopy tools cannot be applied directly to MCNTs because of multiple absorptions and emissions between concentric cylindrical structures. We note that, in the literature, there are very few infrared investigations on SCNTs and on MCNTs. We observed two regions, where vibrational structure is apparent: around 1200 and 1584 cm^{-1}[22]. The first band at 1200 cm^{-1} arises from the translational symmetry of the diamond lattice [23]. The second one has been appeared in the range of the E_{1u} infrared active-modes of graphite at 1590cm^{-1} [24].

Many works on the phonons in MCNTs using the Raman spectroscopy have been devoted [25-27]. In this context, the Raman spectra of infinite MCNTs have been calculated in a previous works [26, 27]. To have a deeper insight on the vibrational properties of MCNT, the aim of this study is to identify the IR active-modes in MCNT.

In this paper, the IR calculated spectra of MCNTs in the breathing-like Mode (BLM), intermediate-like mode (ILM) and tangential-like mode (TLM) ranges were investigated. We report the dependence of wavenumber on walls number in MCNT and on the innermost tube diameter. The main aim of the present paper is to use our results in the interpretation of experimental ir features of MCNTs.

2. Models and Computational Methods

A MCNT consists of concentric single-walled carbon nanotubes (SCNTs) greatly spaced by intermolecular distance d=0.34 nm close to the interlayer distance in graphite. In MCNTs, the relation between the diameters, D_1 and D_N, of the inner and outer tubes is $D_N = D_1 + 2(N-1)d$, where N is the number of walls tube forming the MCNT. The C-C interactions between the carbon atoms of each layer are described using the force constants model published by Saito [28] and previously employed to simulate the IR and Raman spectra of SCNTs [9,11,26]. The inter-tube interactions are described by the Lennard-Jones potential, given by the following expression:

$$U_{LJ}(r) = 4\varepsilon[(\sigma/r)^{12} - (\sigma/r)^{6}] \quad (1)$$

The parameters values of the Lennard-Jones were chosen as $\varepsilon = 2.964$meV and $\sigma = 0.3407$nm. The IR absorption intensity of the spectrum is given by [29]:

$$I_\alpha(\omega) = \frac{\omega}{nc}\sum_j \frac{|a_{j\alpha}|}{2\omega_j}\left(\delta(\omega-\omega_j) - \delta(\omega-\omega_j)\right) \quad (2)$$

With

$$a_{j\alpha} = \sum_k \frac{q_k}{\sqrt{M_k}} e_j(k\alpha) \quad (3)$$

c and n are respectively the speed of light and the refractive index of the material. q_k is the effective charge tensor and M_k the mass of the kth atom. $e_j(k\alpha)$ and ω_j are the ($k\alpha$) component of the displacement amplitude and the wave number for the kth atom (α is Cartesian coordinate) in the j^{th}mode. A usual method to simulate the IR spectra consists of injecting in the previous expressions the values of ω_j and $e_j(k\alpha)$ obtained by direct diagonalization of the dynamical matrix $\tilde{D}$ of the system by resolving the equation

$$\tilde{D}\,|j\rangle = \omega_j^2\,|j\rangle \quad (4)$$

To enhance the IR response of the systems, dynamical effective charges on tied carbon atoms were fixed at $q_A = +1$ and $q_B =- 1$ for a given A-B bond.

3. Results and Discussion

In this section, we focus on the calculated IR lines for infinite MCNTs, which are obtained by applying periodic conditions. In order to identify all IR active modes and their symmetries, we present in Figure 1 the ZZ (dashed lines) and XX (solid lines) calculated IR spectra for typical zigzag MCNTs, as a function of the number of the single tube in individual MCNT N=3, 4, 5, and 6. The spectra are displayed in the Breathing-like Modes (BLM) (right), intermediate modes (middle) and tangential-like modes regions (TLM) (left).

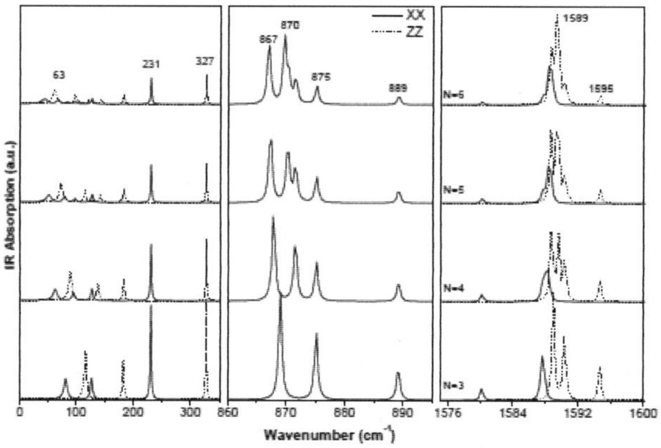

Figure 1. The ZZ (dashed lines) and XX (solid lines) calculated Infrared spectra for typical zigzag (9,0) MCNTs (Din=0,7nm), as a function of the number of walls in individual MCNTs N=3, 4, 5, and 6. The left panel is the BLM region, the middle is the intermediate region and the right panel is the TLM one.

The results for a typical zigzag MCNTs show that, in TLM region, spectra present many peaks, located between 1580 and 1595 cm^{-1}, corresponding to active modes for each individual tube of MCNTs. When the number of tubes increases, this region is dominated by a strong peak around 1589 cm^{-1} for both polarizations. As a consequence, a strong band located in the TLM region is expected to be experimentally observed. More generally, since the sample is a multilayer tubes of different sizes, formed of tubes for different diameters and chirality, a broadening of the TLM band is expected in the IR spectra measured for MCNTs macroscopic samples.

In the intermediate wavenumber (ILM) range, for a given number of tubes N of MCNTs, we found that the infrared spectra present N modes for XX polarization, located between 865 and 890 cm^{-1}. Independently of the number of layers, no infrared-active mode appears in the ZZ polarization. A dominant mode located at 870 cm^{-1} is predicted, independently of the number of layers. In a real sample, the number of tubes in a multilayer tube varies from a few tubes to ten of tubes. As a consequence, a broad band located in the intermediate range is expected in the experimental infrared spectrum of a MCNTs macroscopic sample.

In the BLM region, we note the same behaviour of active modes in the intermediate one. The spectrum presents N modes for both polarizations, for each individual tube forming the system. In this region, also we can note that the vibration of the most internal and external layers are the intense modes. However, when the number of tubes increases, their intensities decrease.

To illustrate the in-phase and the counter-phase collective motions of inner and outer tubes, we plot in Figure 2 the eigenvector displacements of the E_1 BLM

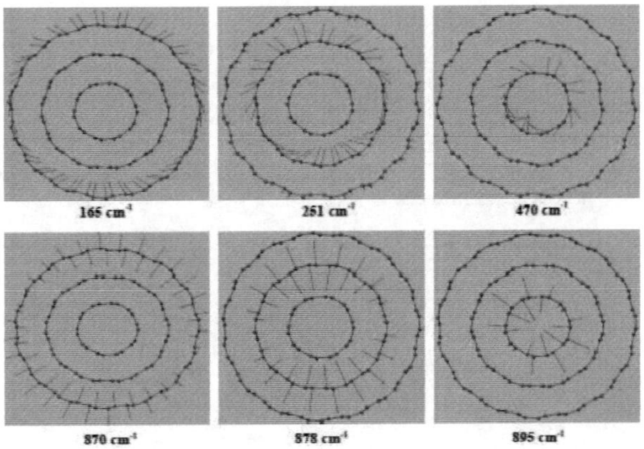

Figure 2.BLM and ILM atomic displacements for selected E1 modes in the (5,5)@(10,10)@(15,15) 3-CNT. The arrows indicate the magnitude and direction of C atom displacements

and ILM for (5,5)@(10,10)@(15,15) triple-walled nanotubes (3-CNT). These vectors allow us to identify the mixing of tangential and radial character of these optical modes: (165,895), (251,878) and (470,870) cm^{-1} for in phase, mixed-phase and counter-phase modes, respectively. From these results, we conclude that the inner tube diameter and the number of layers in the MCNTs are two parameters of primary importance. In comparison with SCNTs Raman spectra, we note that the BLM region of MCNTs is characterized by an upshift of the frequency modes. The ILM and TLM modes regions of SCNTs are slightly affected by MCNTs.

We report in Figure 3 (right), the XX polarized calculated infrared spectra for the zigzag MCNTs in the BLM region for three innermost diameter tubes D_{in} = 0.7, 1.22, and 2.5 nm. The number of layers were fixed to N=15 tubes. In the left of Figure 3 spectra are displayed for four values of N=5, 10, 15, and 18 with innermost diameter fixed to D_{in} = 0:7 nm. We note at highest wavenumbers (Figure 3 on the right) that the modes originate from the inner tube and the intensity of these modes decreases when the inner diameter wall increases. As we can see these modes are too weak to be easily observed for MCNTs with large inner diameter wall. Concerning the MCNTs spectra dependence with N walls (Figure 3 on the left), with fixed inner diameter, the two modes at highest wavenumbers are strongly sensitive to the number of walls. The intensity of all modes decreases when the number of tubes increases.

To illustrate the intermediate wavenumber dependence on the number of walls in MCNTs and on the innermost tube diameter, we have presented the XX polarized infrared spectra for a typical zigzag MCNTs for four numbers of tubes N=5, 10, 15, and 18 with innermost diameter fixed to D_{in} = 0.7 nm (Figure 4: left). In the left of Figure 4, IR spectra are displayed for three innermost diameter tubes D_{in} = 0.7, 1.22, and 2.5 nm, with a number of layers fixed to N=15 tubes.

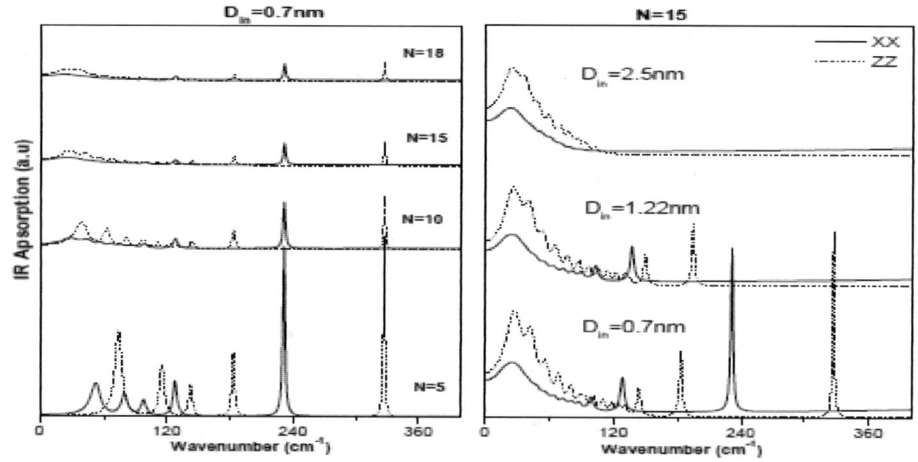

Figure 3. The ZZ (dash dot lines) and XX (solid lines) calculated Infrared spectra for typical zigzag MCNTs in the BLM region, upon the number of walls in individual MCNTs N=5, 10, 15, and 18. The left panel indicates the effect of the number of walls on MCNTs of inner diameter D_{in} = 0.7nm, whose number of walls varies between 5 and 18. The right panel shows the effect of the inner diameter D_{in} on 15 wall tubes.

We found that for large innermost diameters and when the number of walls increases, one can observe that the intermediate infrared peaks progressively change in intensity and overlap leading to a single band centred at 870 cm^{-1} close to the out-of-plane radial buckling mode in graphite.

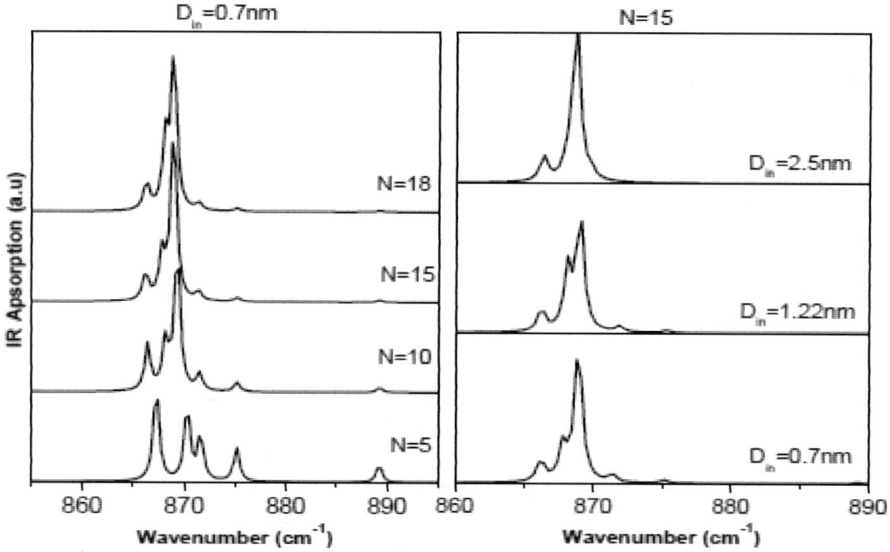

Figure 4. The ZZ calculated Infrared spectra for typical zigzag MCNTs in the ILM region, upon the number of walls in individual MCNTs N=5, 10, 15, and 18. The left panel indicates the effect of the number of walls on MCNTs of diameter D_{in} = 0.7nm, whose number of walls varies between 5 and 18. The right panel shows the effect of the inner diameter D_{in} on 15 wall tubes.

Now, we focus on the dependence of the polarized infrared spectrum as a function of the number of walls. Figure 5 shows the evolution of the position of the wavenumber of the intermediate modes in zigzag MCNTs as a function of the number of walls. As expected we note that the number of infrared modes increases with increasing the size of the MCNTs. For a given inner diameter D_{in} = 0.7 nm, the wavenumber of the lowest mode decreases monotonically to 800 cm^{-1}, as the number of layers increased, while the wavenumber of the higher mode tends to a constant value of about 894 cm^{-1}.

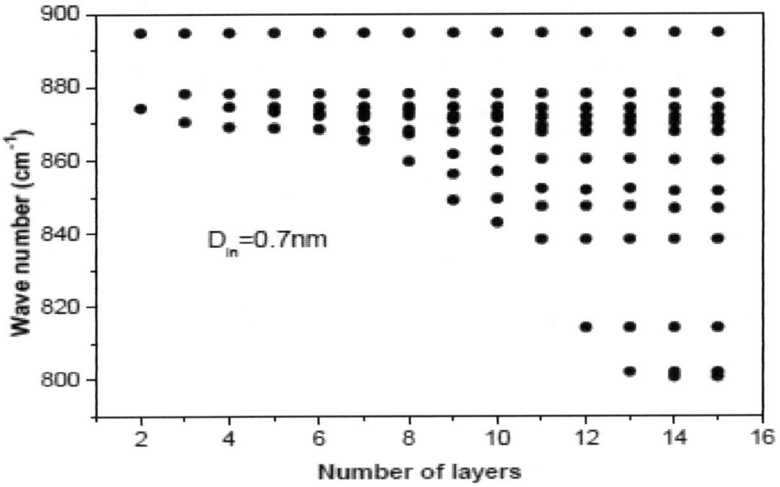

Figure 5. Dependence of the wavenumbers of the intermediate wavenumber modes of multilayer tubes of zigzag MCNTs, calculated for a fixed diameter Din = 0.7nm of the innermost layer and different numbers of layers N.

4. Conclusion

In this article, the infrared spectra of MCNTs in large diameters are reported taking advantage of the spectral moment's method. The dependence of the positions of the infrared-active modes in the low and intermediate-wavenumber ranges as a function of the number of walls in MCNTS is discussed. We find that for large number of walls, the position and intensity of all intermediate modes overlap to give a large band located at 870 cm^{-1} close to the out-of-plane radial buckling mode in graphite. Our method is well adapted to understand the experimental infrared spectra of MCNTs.

References

[1] Iijima S 1991 *nature.* **354** 56
[2] Bethune D, Kiang C H, De Vries M, Gorman G, Savoy R, Vazquez J and Beyers R 1993 *Nature*. **363** 605
[3] Scheibe B, Borowiak-Palen E and Kalenczuk R 2010 *J. Alloy. Compd.* **500** 117
[4] Yang C, Wang D, Hu X, Dai C and Zhang L 2008 *J. Alloy Compd.* **448** 109
[5] Endo M, Kim C, Nishimura K, Fujino T and Miyashita K 2000 *Carbon.* **38** 183
[6] Lund P 1991 *Int. J. Hydrogen. Energ.* **16** 735
[7] Tu W C, Chang Y T, Yang C H, Huang C T, Yeh D J, Ho C I, Hsueh C Y and Lee S C 2011 IEEE. 680-683
[8] Kim D Y, Yang C M, Park Y S, Kim K K, Jeong S Y, Han J H and Lee Y H 2005 *Chem. Phys. Lett.* **413** 135
[9] Rahmani A, Sauvajol J L, Cambedouzou J and Benoit C 2005 *Phys. Rev. B.* **71** 125402
[10] Levshov D, Michel T, Paillet M, Than X T, Tran H N, Arenal R, Rahmani A, Boutahir M, Zahab A A and Sauvajol J L 2014 *MRS Online Proceedings Library Archive.* **1700** 69

[11] Roh J Y, Kim Y H and Lee C S 2011 *Curr. Appl. Phys.* **11** S69-S72
[12] Rahmani A, Sauvajol J L, Rols S and Benoit C 2002 *Phys. Rev. B.* **66** 125404
[13] Eatemadi A, Daraee H, Karimkhanloo H, Kouhi M, Zarghami N, Akbarzadeh A, Abasi M, Hanifehpour Y and Joo S.W 2014 *Nanoscale Res. Lett.* **9** 393
[14] Chen C H, Liang Y H and Zhang W D 2010 *J. Alloy. Compd.* **501** 168
[15] Cheung W, Pontoriero F, Taratula O, Chen A M and He H 2010 *Adv. Drug. Deliv. Rev.* **62** 633
[16] Rahmani A, Boutahir M, El Biyaali A, Fakrach B, Chadli H, Sbai K, Hermet P, Bantignies J L and Rahmani A 2016 *RSC Adv.* **6** 41025
[17] Boutahir M, Rahmani A, Fakrach B, Chadli H and Rahmani A 2016 *Int. J. Hydrog. Energy.* **41** 20874
[18] Fakrach B, Rahmani A, Chadli H, Sbai K, Hermet P and Rahmani A 2015 *J. Phys. Chem. C.* **119** 13306
[19] Boutahir M, Rahmani A H, Chadli H and Rahmani A 2016 *Eur. Phys. J-Appl. Phys.* **74** 24605
[20] Sbai K, Rahmani A, Fakrach B, Chadli H and Benhamou M 2014 *Physica E.* **56** 312
[21] Qian W, Liu T, Wei F and Yuan H 2003 *Carbon.* **9** 1851
[22] Kouklin N, Tzolov M, Straus D, Yin A and Xu 2004 *J Appl. Phys. Lett.* **85** 4463
[23] Morelli D T and Uher C 1993 *Appl. Phys. Lett.* **63** 165
[24] Jeon G S and Mahan G 2005 *Phys. Rev. B.* **71** 184306
[25] Zhao X, Ando Y, Qin L C, Kataura H, Maniwa Y and Saito R 2002 *Appl. Phys. Lett.* **81** 2550
[26] Thomsen C 2000 *Phys. Rev. B* **61** 4542
[27] Sbai K, Rahmani A, Chadli H and Sauvajol J L 2008 *J. Phys. Condens. Matter.* **21** 045302
[28] Saito R, Takeya T, Kimura T, Dresselhaus G and Dresselhaus M 1998 *Phys. Rev. B.* **57** 4145
[29] Maradudin A A, Montroll E W, Weiss G H and Ipatova I 1963 *Theory of lattice dynamics in the harmonic approximation.* **3** (Academic press New York,)

The effect of Tri- and Tetra-vacancies defects on the electronic and vibrational properties of (14, 5) chiral carbon nanotube

SA Ait Abdelkader, M Boutahir, AH Rahmani, B Fakrach, M Bentaleb, H Chadli and A Rahmani

Advanced Material and Applications Laboratory (LEM2A), University Moulay Ismail,FSM-ESTM-FPE, BP 11201, Zitoune, 50000 Meknes, Morocco.

E-mail: a.rahmani@fs-umi.ac.ma

Abstract. In this theoretical work, using the spectral moment's method and the bond polarisability model, the non-resonant Raman spectra of (14,5) chiral carbon nanotube with Tri- and Tetra-vacancies defects were calculated. The evolution of the Raman lines of defectives CNTs as a function of the defects concentration was discussed. It is observed that, the Raman spectra of defective carbon nanotubes present the D-band around 1350 cm-1. The Raman characteristics modes of TriV and TetraV defects are identified. The evolution of the electronic density of states as a function of the defects concentration was investigated. This study can provide benchmark to understand the experimental data of defective CNTs.

1. Introduction

Carbon nanotubes [1] and graphene [2] have been extensively studied [3, 4, 5, 6]] given to their exceptional structural, mechanical and electronic properties [7, 8, 9, 10, 11], and are widely used in several applications including electronics, charge transport and biomedical sensing [12, 13]. In practice, it is almost impossible to work with defect-free materials, and some structural point's defects can be introduced unintentionally during the growth [14] or under electron irradiation [15] and chemical treatments. Since the synthetized materials are usually not ideal crystal, the properties mentioned above can be significantly altered due to some unavoidable structural point's defects during the synthesis processes. Several kinds of points defects can be sited in the wall of the synthetized carbon nanotubes [16, 17], such as Stone-Wales defects [18, 19], ad-atoms [20, 21] and vacancies [22, 23, 24]. Among them, the vacancies are the most prolific and important defects, because they certainly have an influence on the mechanical, physical and chemical properties of carbon nanotubes. Therefore, it is very important to understand how the vacancies defects tailor the intrinsic properties of this system. Great attention has been focused on the study of atomic vacancies effect on the carbon nanotube properties [25, 26, 27]. It has been found that the tensile strength and critical stress of carbon nanotubes decrease due to the mono-vacancies and small vacancy clusters [28, 29]. The dangling bonds generated by the vacancies in CNTs (Carbon Nanotubes) can act as active sites for adsorptions [30, 31]. Concerning the vacancies effect on the electronic properties of carbon nanotubes, it's found that the vacancies induced changes and create new available states at around Fermi level [32]. It's estimated that the density of states around the Fermi level rises due to the broken ? bands at the vacancies site [33]. Raman spectroscopy is a characterization technique extremely sensitive to molecules geometries and the bondings between their atoms. This sensitivity is extremely useful for the study of the different carbon allotropes (i.e. carbon nanotubes, fullerenes, graphene, etc.). The Raman spectrum of perfect carbon nanotube is dominated by the so-called RBM (Radial Breathing Mode) band below 350 cm^{-1} and by the G band in the high wave number region. It was found that the RBM mode directly related to the tube diameter [34], and it can be used to estimate the distribution of tube diameters in samples [35]. The G-band is due to the vibration of carbon atoms in plane involving the tube wall. Usually, another mode is observed at around 1350 cm^{-1}. This mode is known by the D-band and is a defect signature in the graphene lattice and is often used to characterize defective carbon nanotubes. This paper discusses the effect of Tri- and Tetra-vacancies defect on the vibrational and electronic properties of (14,5) chiral carbon nanotube and provides an updated overview of the current characterization tools able to identify and detect vacancies-defects in carbon nanotubes.

2. Computational method

We focus on the electronic and the vibrational properties of Tri- and Tetra-vacancies defect in (14,5) chiral carbon nanotube (Figure 1). We use the Raman spectroscopy, such as a perfect tool to get important information and characterization of both perfect and defective carbon nanotube.

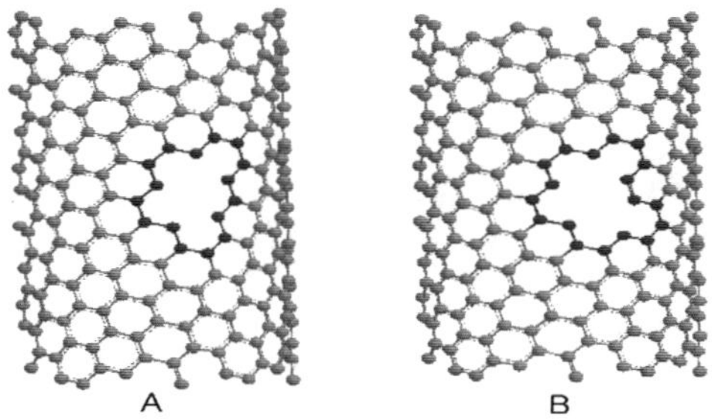

Figure 1. Molecular models of carbon nanotube with Tri- vacancy defect (A) and Tetra-vacancy defects (B).

For this purpose, by using the force constant model introduced by Wirtz and Rubio [36], we described the C-C interactions of the carbon atoms sited on the perfect area between vacancies defects. The interactions C-C of carbon atoms surounding the vacancies defects are calculated by using Density Functional Theory (DFT) as implemented inside the SIESTA package [**?**].

The Raman intensity is obtained according to the bond polarizability model [**?**]. In this approach, the polarization is just modulated for the nearest-neighbor bonds and the components of the induced polarizability tensor are given by the empiric equation:

$$\pi_{\alpha\beta}(r) = \frac{1}{3}(\alpha_l + 2\alpha_p)\delta_{\alpha\beta} + (\alpha_l - \alpha_p)\left(\hat{r_\alpha}\hat{r_\beta} - \frac{1}{3}\delta_{\alpha\beta}\right) \tag{1}$$

where α and β relate to the cartesian components (x,y,z) and $\overrightarrow{r}$ is the unit vector along the vector $\vec{r}$ connecting atom n and atom m which are covalently bonded. The parameters α_l and α_p correspond to the longitudinal and perpendicular bond polarizability respectively. In this approximation, one can assume that the bond polarizability parameters are depending on the bond lengths r only. The derivatives $\pi^n_{\alpha\beta,\gamma}$ are given by:

$$\pi^n_{\alpha\beta,\gamma} = \sum_m \frac{1}{3}(\alpha_l' + 2\alpha_p')\delta_{\alpha\beta}\hat{r_\gamma} + (\alpha_l' - \alpha_p')\left(\hat{r_\alpha}\hat{r_\beta} - \frac{1}{3}\delta_{\alpha\beta}\right)\hat{r_\gamma} + \frac{\alpha_l - \alpha_p}{r}(\delta_{\alpha\gamma}\hat{r_\beta} + \delta_{\beta\gamma}\hat{r_\alpha} - 2\hat{r_\alpha}\hat{r_\beta}\hat{r_\gamma}) \quad (2)$$

where $\alpha' = \left(\frac{\partial\alpha}{\partial r}\right)\Big|_{r=r_0}$ and r_0 is the equilibrium bond distance. The bond polarizability parameters are:

$$\overline{\alpha} = 2\alpha_p' + \alpha_l', \quad \overline{\beta} = \alpha_l' + \alpha_p', \quad and \quad \overline{\gamma} = \frac{\alpha_l - \alpha_p}{r} \quad (3)$$

3. Results and discussion

In this section, the effect of Tri- (Tetra-) vacancies on the vibrational and electronic properties of (14,5) chiral carbon nanotube is investigated. The defect concentration is defined as the report of the number of the removed atoms by the total number of atoms in the perfect tube.

3.1. Raman Spectra of Carbon Nanotubes with Tri- And Tetra-Vacancies Defects

We report the evolution of the Raman spectra of the defective (14,5) carbon nanotube as a function of the Tri- and Tetra-vacancies defect concentration. The vacancies concentration varied from 0% (for perfect tube) 1%, 2%, 3%, 4% and 5% (for the defective carbon nanotube). The studied samples are composed by a large number of carbon atoms (up to 24000 degrees of freedom). The maximum number of Tri- (Tetra-) vacancies defects can reach 1332 (1000) respectively. The nanotube axis is along the Z-axis. The laser beam is kept along the Y-axis of the reference frame. We consider that both incident and scattered polarizations are along the Z-axis to calculate the polarized ZZ spectra. The evolution of the obtained Raman spectra as function of tri-vacancies defect concentration are reported in the panels of Figure 2 with their corresponding pure nanotube (indicated by 0%). The spectra are displayed in the RBM (Radial Breathing Mode) range (left), intermediate range (middle), and the tangential modes range (right).

It is observed from the figure 2, that there is a significant effect of the Tri-vacancies defects on the Raman lines of the perfect tube. As we can see, the frequency of the RBM mode decreases as a function of the tri-vacancies concentration. When the concentration of tri-vacancies increases, the RBM downshifts from 166 cm^{-1} in pristine tube to 164 cm^{-1}, 163 cm^{-1}, 161 cm^{-1}, 159 cm^{-1}, 158 cm^{-1} for 1%, 2%, 3%, 4% and 5% of tri-vacancies concentration on the (14,5) SWNT (Single-Walled Carbon nanotubes), respectively. A second-lowest Raman-active peak located at around the value 450 cm^{-1} is observed in the case of defect tube only. In the intermediate range, the spectra of defectives nanotubes have shown the D band at around 1340 cm^{-1}, while no Raman line is expected for the perfect tube. This mode arises from tri-vacancies defect and its intensity depends on the defect concentration. Similarly to the previous calculations for SWNTs, the G-band located in the TM range is weakly downshift as a function of defect concentration. For instance, the G mode downshifts from 1591 cm^{-1} in the perfect nanotube to 1586 cm^{-1} in the 5% defective nanotube. The eigenvector displacement of the mode located around 450 cm^{-1} in spectra of SWNTs with Tri-vacancies defect, is given in Figure (3-A). These vectors allow us to identify the radial displacements of atoms belonging to the Tri-vacancy ring. This mode could be used as the intrinsic Raman signature of the Tri-vacancies defect sited in the carbon nanotube.

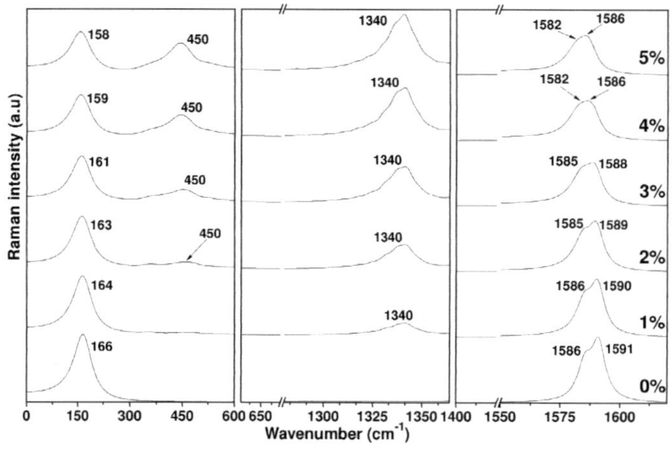

Figure 2. Evolution of the Raman spectra of (14,5) nanotube as a function of Tri-vacancies defects concentration.

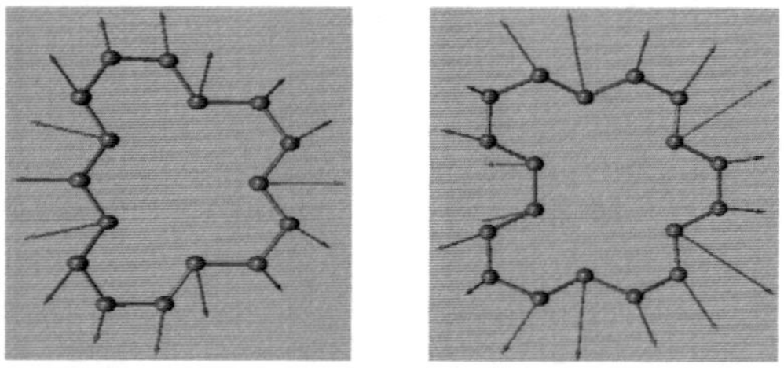

Figure 3. In A and B, the atomic displacements of Tri-vacancy and Tetra-vacancy in (14,5) nanotube, due to the Raman active modes at 450 cm^{-1} and 420 cm^{-1} respectively.

The Raman spectra of the chiral carbon nanotube (14.5) with Ttra-vacancy defects were calculated. The evolution of the obtained spectra as a function of the concentration of defects is illustrated in figure 4. As shown in the figure 4, the G mode (panel in the right) is slightly shifted to the low frequencies when the defect concentration increases. For example, the frequency of the G mode varies from 1591 cm^{-1} to the perfect nanotube (indicated by 0%) at 1586 cm^{-1} for the tube with 5% of Tetra-vacancies defects. In the intermediate zone, the spectra of the defective tubes have the D band around 1330 cm^{-1}. The eigenvector displacements of the mode around 420 cm^{-1} in the spectra of nanotubes with Tetra-vacancies defects are given in the figure (3-B). These vectors allow us to identify the radial displacements of atoms belonging to the Tetra-vacancy defect. In this mode, the atoms constituting the Tetra-vacancy ring moves in phase in the radial direction. This mode could be used as the Raman signature of the Tetra-vacancy sited in the carbon nanotube.

3.2. Band Structures and Density of States of Carbon Nanotubes with Tri- And Tetra-Vacancies Defects

Using the tight-binding model, the electronic bands structures and the electronic density of states of pristine and defective (14,5) carbon nanotube were calculated. These calculations were carried out for one unit cell of (14.5) tube (for pristine cell and defective cell with one and two

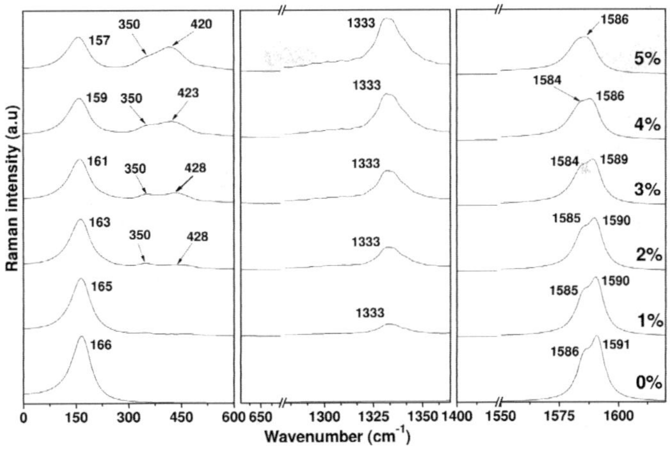

Figure 4. Evolution of the Raman spectra of (14,5) nanotube as a function of Tetra-vacancies defects concentration.

vacancies). The periodic conditions are applied along the axis of the tube. We present the obtained results in the direction ...M of the irreducible Brillouin zone of the structure of our system. The obtained bands structures and the densities of states of the (14, 5) tube with Tri- and Tetra-vacancies defects are shown in figures 5 and 6 respectively.

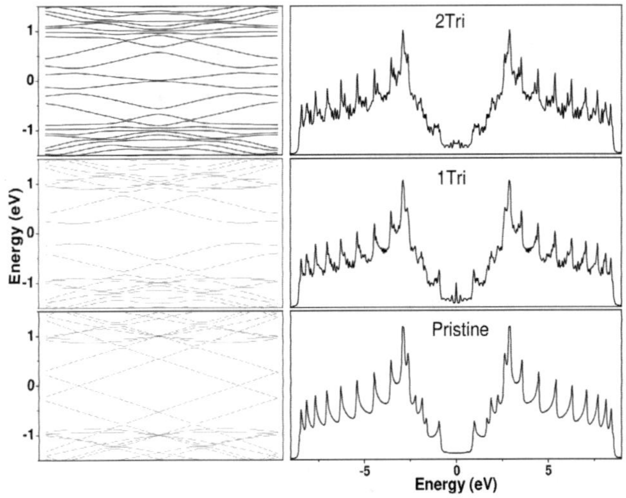

Figure 5. Band structure and density of states for (14,5) nanotube with Tri-vacancies defects.

In the case of Tri-vacancy defect in the (14.5) tube (Figure 5), we observe that there are new electronic states around the Fermi level. These electronics states are indicated by three peaks located around the Fermi level on the density of states of the tube with single Tri-vacancy (indicated by 1Tri). The first one is located at Fermi level (EF = 0 eV) and the two others are symmetric with respect to Fermi level. For the tube with two Tri-vacancies (2Tri), we observe that there are several new adjacent peaks. Concerning the Tetra-vacancy, we observe also that

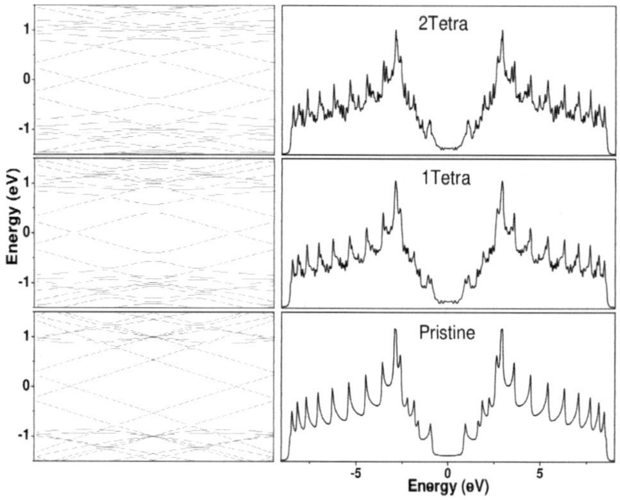

Figure 6. Band structure and density of states for (14,5) nanotube with Tetra-vacancies defects.

there are several new electronic states between the valence band and the conduction band. These new electronics states appeared like a several small peaks for the tube with one vacancy (two vacancies) as indicated respectively by 1Tetra (2Tetra) in figure 6. Such new electronics states could be due to the dangling bonds sited in the vacancies that introduced the localized empty defect states between the conduction band and valence band of the nanotube.

4. Conclusion

In this paper, the non-polarized Raman spectra of chiral (14,5) single-walled carbon nanotube with Tri- and Tetra-vacancies defects as a function of such defects concentration were investigated. Characteristic peaks of Tri- and Tetra-vacancies defects are identified in the radial breathing modes region. The obtained D band is in agreement with the experimental measurements. Using the tight-binding model, the electronic band structures and the density of states of (14,5) carbon nanotubes with the Tri- and Tetra-vacancies were also calculated. For the defective tubes, new electronic states around the Fermi level are identified.

Acknowledgments
The works was supported by a CNRST-Morocco agreement.

References

[1] S. Iijima 1991 Helical microtubules of graphitic carbon, nature 354 (6348) 56.
[2] K. S. Novoselov, A. K. Geim, S. V. Morozov, D. Jiang, Y. Zhang, S. V. Dubonos, I. V. Grigorieva, A. A. Firsov 2004 Electric field effect in atomically thin carbon films, science 306 (5696) 666-669.
[3] M. Boutahir, A. Rahmani, H. Chadli, A. Rahmani 2016 Mechanical coupled vibrations in an individual double-walled carbon nanotube, The European Physical Journal Applied Physics 74 (2) 24605.
[4] T. W. Odom, J.-L. Huang, P. Kim, C. M. Lieber 1998 Atomic structure and electronic properties of single-walled carbon nanotubes, Nature 391 (6662) 165 62.
[5] J. Mintmire, C. White 1995 Electronic and structural properties of carbon nanotubes, Carbon 33 (7) 893-902.
[6] M. Boutahir, A. Rahmani, B. Fakrach, H. Chadli, A. Rahmani 2016 Theoretical study of electronic and vibrational properties of dimer of single-wall carbon nanotubes, international journal of hydrogen energy 41 (45) 20874-20879

[7] I. Frank, D. M. Tanenbaum, A. M. van der Zande, P. L. McEuen 2007 Mechanical properties of suspended graphene sheets, Journal of Vacuum Science and Technology B: Microelectronics and Nanometer Structures Processing, Measurement, and Phenomena 25 (6) 2558-2561.
[8] S.-M. Choi, S.-H. Jhi, Y.-W. Son 2010 Effects of strain on electronic properties of graphene, Physical Review B 81 (8) 081407.
[9] J. N. Coleman, U. Khan, W. J. Blau, Y. K. Gunko 2006 Small but strong: a review of the mechanical properties of carbon nanotubepolymer composites, Carbon 44 (9) 1624-1652.
[10] T. Ando 2009 The electronic properties of graphene and carbon nanotubes, NPG asia materials 1 (1) 17.
[11] A. Jorio, G. Dresselhaus, M. S. Dresselhaus 2007 Carbon nanotubes: advanced topics in the synthesis, structure, properties and applications, Springer Science and Business Media Vol. 111, 185.
[12] G. Dresselhaus, S. Riichiro, et al. 1998 Physical properties of carbon nanotubes, World scientific.
[13] D. B. Mawhinney, V. Naumenko, A. Kuznetsova, J. T. Yates Jr, J. Liu, R. Smalley 2000 Surface defect site density on single walled carbon nanotubes 190 by titration, Chemical Physics Letters 324 (1-3) 213-216.
[14] J. A. Rodriguez-Manzo, F. Banhart 2009 Creation of individual vacancies in carbon nanotubes by using an electron beam of a diameter, Nano letters 9 (6) 2285-2289.
[15] T. Ebbesen, T. Takada 1995 Topological and sp3 defect structures in nanotubes, Carbon 195 33(7) 973-978.
[16] J.-C. Charlier 2002 Defects in carbon nanotubes, Accounts of chemical research 35 (12) 1063-1069.
[17] H. F. Bettinger 2005 The reactivity of defects at the sidewalls of single-walled carbon nanotubes: The stone-wales defect, The Journal of Physical Chemistry B 109 (15) 6922-6924.
[18] L. Zhou, S.-Q. Shi 2003 Formation energy of stone-wales defects in carbon nanotubes, Applied physics letters 83 (6) 1222-1224.
[19] P. G. Collins 2010 Defects and disorder in carbon nanotubes, Oxford Handbook of Nanoscience and Technology: Materials: Structures, Properties 205 and Characterization. Techniques 2 (2) 31.
[20] P. Lehtinen, A. S. Foster, A. Ayuela, A. Krasheninnikov, K. Nordlund, R. M. Nieminen 2003 Magnetic properties and diffusion of adatoms on a graphene sheet, Physical review letters 91 (1) 017202.
[21] S. Zhang, S. L. Mielke, R. Khare, D. Troya, R. S. Ruo, G. C. Schatz, 210 T. Belytschko 2005 Mechanics of defects in carbon nanotubes: atomistic and multiscale simulations, Physical Review B 71 (11) 115403.
[22] A. Hashimoto, K. Suenaga, A. Gloter, K. Urita, S. Iijima 2004 Direct evidence for atomic defects in graphene layers, Nature 430 (7002) 870.
[23] R. G. Amorim, A. Fazzio, A. Antonelli, F. D. Novaes, A. J. da Silva, 2007 Divacancies in graphene and carbon nanotubes, Nano Letters 7 (8) 2459-2462.
[24] M. Huhtala, A. Krasheninnikov, J. Aittoniemi, S. Stuart, K. Nordlund, K. Kaski 2004 Improved mechanical load transfer between shells of multiwalled carbon nanotubes, Physical Review B 70 (4) 045404.
[25] S. A. A. Abdelkader, F. Fergani, M. Bentaleb, M. Boutahir, B. Fakrach, H. Chadli, A. Rahmani 2016 Trivacancy defects and their effects on the electronic and vibrational properties of single-walled carbon nanotubes, in: 2016 International Renewable and Sustainable Energy Conference (IRSEC), IEEE 50-53 (225) 9
[26] J. Yuan, K. M. Liew 2009 Effects of vacancy defect reconstruction on the elastic properties of carbon nanotubes, Carbon 47 (6) 1526-1533.
[27] M. Sammalkorpi, A. Krasheninnikov, A. Kuronen, K. Nordlund, K. Kaski 2004 Mechanical properties of carbon nanotubes with vacancies and related defects, Physical Review B 70 (24) 245416.
[28] M. Terrones, H. Terrones, F. Banhart, J.-C. Charlier, P. Ajayan 2000 Coalescence of single-walled carbon nanotubes, Science 288 (5469) 1226-1229.
[29] H. He, B. Pan 2008 Electronic structures and vibrational properties of a carbon nanotube with adsorption of small hydrocarbon radicals, The Journal of Physical Chemistry C 112 (48) 18876-18881.
[30] N. Neophytou, S. Ahmed, G. Klimeck 2007 Influence of vacancies on metallic nanotube transport properties, Applied physics letters 90 (18) 182119.
[31] E. Faizabadi 2011 Single wall carbon nanotubes in the presence of vacancies and related energy gaps, in: Electronic Properties of Carbon Nanotubes, IntechOpen.
[32] M. S. Dresselhaus, G. Dresselhaus, R. Saito, A. Jorio 2005 Raman spectroscopy of carbon nanotubes, Physics reports 409 (2) 47-99.
[33] S. Rols, A. Righi, L. Alvarez, E. Anglaret, R. Almairac, C. Journet, P. Bernier, J. Sauvajol, A. Benito, W. Maser, et al. 2000 Diameter distribution of single wall carbon nanotubes in nanobundles, The European Physical Journal B-Condensed Matter and Complex Systems 18 (2) 201-205.
[34] L. Wirtz, A. Rubio 2004 The phonon dispersion of graphite revisited, Solid State Communications 131 (3-4) 141-152.
[35] J. M. Soler, E. Artacho, J. D. Gale, A. Garcia, J. Junquera, P. Ordejon, D. Sanchez-Portal 2002 The siesta method for ab initio order-n materials simulation, Journal of Physics: Condensed Matter 14 (11) 2745.
[36] S. Guha, J. Menendez, J. Page, G. Adams 1996 Empirical bond polarizability model for infullerenes, Physical

Review B 53 (19) 13106.

IOP Conf. Series: Materials Science and Engineering **783** (2020) 012019 doi:10.1088/1757-899X/783/1/012019

Dosimetric investigation for the new BEBIG cobalt 60 source used in brachytherapy using Monte Carlo N-Particles eXtended code (MCNPX)

S Elboukhari [1*], K Yamni [1], H Ouabi [2], T Bouassa [3] and L A Mlouk [4]

[1] EMaMePS, Faculty of Sciences, My Ismail University, Morocco.
[2] Center of Oncology ALAZHAR, Morocco.
[3] Faculty of Sciences, Mohammed V University, Rabat, Morocco.
[4] Faculty of Sciences, My Ismail University, Meknes, Morocco.

[*]Corresponding author: s.elboukhari@edu.umi.ac.ma

Abstract: The purpose of this work is to calculate the parameters of the ^{60}Co-source model: Co0.A86 newly introduced to the HDR brachytherapy, and to validate the MCNPX code in Monte Carlo investigation, the radial dose function, and function of anisotropy are calculated in water as recommended in the report of HEBD working group (AAPM&ESTRO). The results obtained in this study agree will with the data published in previous studies.

1. Introduction

Brachytherapy is a local irradiation. The dose is delivered by one or several sealed sources. Different kinds of brachytherapy have been defined. Nowadays, Ir-192 and Co-60 are available in devices used in high dose rate HDR brachytherapy, generally for the treatment of gynecological tumors. The dose rate distribution in the formalism of HEBD working group (AAPM&ESTRO) requires calculating different dosimetric parameters of the new BEBIG Eckert & Ziegler ^{60}Co HDR source, using a Monte Carlo code. The dosimetric data given in this study can be useful for the control of TPS calculations.

2. Materials and methods

2.1. Source description and geometry

The HDR brachytherapy source of 60 Co, model Co0.A86, manufactured by BEBIG used in this work was composed of pure cobalt 60 cylindrical active core with a 3.5mm length and 0.5mm in diameter (density:8.09 g/cm^3). The active part is covered by an air shell of 0.1 mm of thickness then encapsulated in a cylindrical stainless steel capsule of 0.2mm for thickness and 1mm for the external diameter. For this study, we considered 4mm and 0.9mm for the source cable length and diameter, respectively, figure 1. The density used for the stainless steel, 8.03 g/cm^3 for the capsule and the source cable.

Table 1. Elemental composition used in this study (source and phantom) [1].

Element	Medium		
	Water (%)	Air (%)	Stainless steel (%)
H	11.010	0.073	-
C	-	0.012	0.03
N	-	75.032	0.01
O	88.900	-	-
Si	-	-	0.75
Na	-	-	-
Mg	-	-	-
Cl	-	-	-
K	-	-	-
Ca	-	-	-
P	-	-	0.045
S	-	-	0.03
Ar	-	1.274	-
Cr	-	-	17.0
Mn	-	-	2.0
Fe	-	-	65.543
Ni	-	-	12.0
Mo	-	-	2.5
Co	-	-	-
Density (g/cm^3)	0.998	0.012	8.03

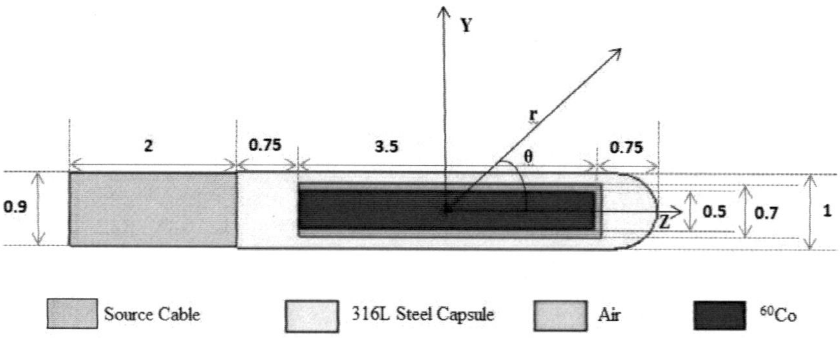

Figure 1. Source description (dimensions in mm)

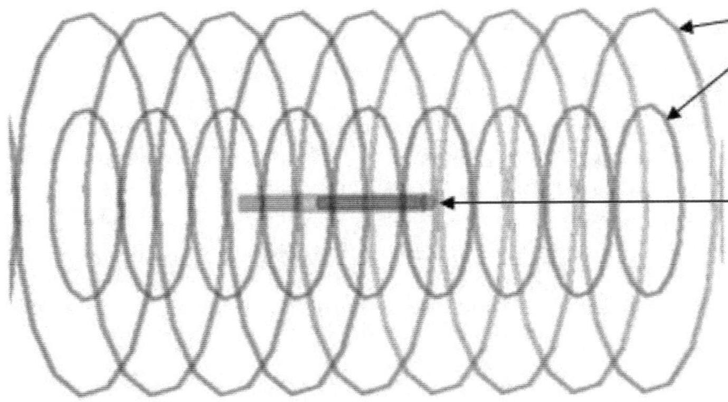

Figure 2. The source and the detectors geometry used to calculate $g_y(r)$ with MCNPX

2.2. Dose calculation formalism

According to the recommendations of the HEBD, the dose formalism for the measure of dose rate can be defined in a polar coordinate system by the equation (1)[2]:

$$\dot{D}(r,\theta) = Sk\,\Lambda\,\frac{G(r,\theta)}{G(r0,\theta 0)}\,g_y(r)\,F(r,\theta) \qquad (1)$$

This formalism required the calculations of dosimetric parameters of the source using Monte Carlo code, such as the air-kerma strength S_k, the radial dose function $g_y(r)$, the anisotropy function $F(r,\theta)$, and the dose rate constant Λ. The radial dose function and the anisotropy function are calculated using the two following equations:

$$g_y(r) = \frac{D(r,\theta 0)}{D(r0,\theta 0)}\,\frac{G(r0,\theta 0)}{G(r,\theta 0)} \qquad (2)$$

$$F(r,\theta) = \frac{D(r,\theta)}{D(r,\theta 0)}\,\frac{G(r,\theta 0)}{G(r,\theta)} \qquad (3)$$

The geometry function $G(r, \theta)$ is calculated using the following expression:

$$G(r,\theta) = \frac{\tan^{-1}\left[\frac{r\cos\theta + \frac{L}{2}}{r\sin\theta}\right] - \tan^{-1}\left[\frac{r\cos\theta - \frac{L}{2}}{r\sin\theta}\right]}{Lr\sin\theta} \quad \text{for } \theta \neq 0 \qquad (4)$$

$$G(r,\theta) = \left(r^2 - \frac{L^2}{4}\right)^{-1} \quad \text{for } \theta = 0 \qquad (5)$$

Where r is the distance to the point of interest and θ is the polar angle with the longitudinal axis of the source (Z axis). The point ($r0$ =1cm, $\theta 0$=90°)) is chosen as a reference located in the transverse axis of the source.

2.3. Monte Carlo calculation.

For this study, we use the MCNPX version: 2.70 (license: C00810MNYCP), with the visual Editor VisedX_24E, this edition includes several tools for geometry modeling, particles transport, and a 3D viewing of the defined geometry of the source. To define the dosimetric parameters, MCNPX has different tallies for each chosen calculation. For the simulation, we consider the source composed of two gamma-rays with energies: 1.17 MeV and 1.332 MeV [3]. The

Monte Carlo code fulfills all the recommendations of the report, "Dosimetric prerequisites for routine clinical use of photon emitting Brachytherapy sources" of the AAPM.

The physics "low energy" models of MCNPX have been used. These physics models use the xdata/mcplib84 cross-sections library for photons and the xdata/el03 for electrons.

The gamma spectrum used in this work was obtained from the database NuDat. A cutoff energy of 10 keV was used. 2×10^9 photon history was simulated to calculate the dose with an Intel® Xeon (R) CPU E5620@2.40GH×16, HP-Z600. The methodology used for the Monte Carlo simulation follows the HEBD report. In order to obtain the radial dose and anisotropy function in water, the source has been located in the center of a spherical water phantom of 40 cm in radius it acts as an unbounded phantom up to a distance of 20 cm from the source center. HEBD recommended liquid water with a density of 0.998 gcm^{-3} at 22 °C. A grid system of cylindrical rings 0.05 cm thick and 0.05 cm high concentric to the longitudinal source axis, was used to obtain the radial dose function and, to obtain the anisotropy function in the form given by the HEBD report a system of 400*180 concentric spherical sections with a variable thick: $(0.1\text{mm})^3$ voxels for $r\leq 1$cm, $(0{,}5\text{mm})^3$ for 1cm< $r \leq$ 5cm, $(1\text{mm})^3$ for 5cm< $r \leq$ 10cm, and $(2\text{mm})^3$for 10cm< $r \leq$ 20cm, with an angular width of 1° in the polar angle was used[4]. The coordinate axes used are shown in Figure1.

Table 2. Radial dose function values calculated by MCNPX.

r (cm)	$g_y(r)$			
	This work	H. Badry et al.	Guerrero et al.	Granero et al.
0.25	1.0143	1.06898	1.0707	1.007
0.5	0.9977	1.02743	1.0291	1.036
0.75	1.0046	1.0151	1.0102	1.015
1	1	1	1	1
1.5	0.9926	0.99312	0.9894	0.992
2	0.9874	0.97814	0.9794	0.984
3	0.9689	0.96728	0.9638	0.968
4	0.9539	0.95048	0.9471	0.952
5	0.9378	0.93074	0.9335	0.939
6	0.9205	0.91969	0.915	0.919
7	0.9035	0.90192	0.898	0.902
8	0.8867	0.88179	0.883	0.884
9	0.8683	0.86489	------	------
10	0.8513	0.84412	0.847	0.849
12	0.8156	0.81113	0.813	0.813
15	0.7593	0.75372	0.756	0.756
20	0.6628	0.66381	0.661	0.665

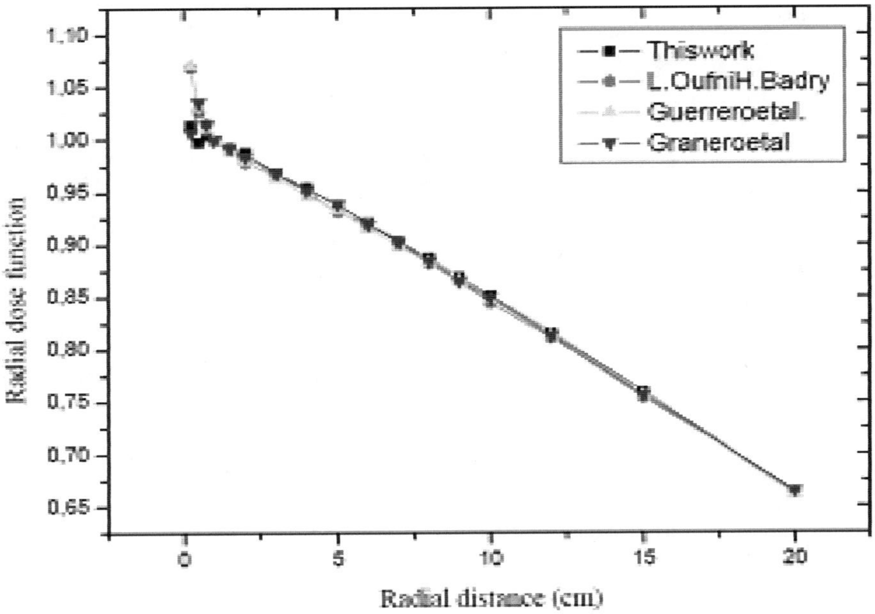

Figure 3. Radial dose function obtained with MCNPX

Table 3. Elemental composition of media used to compare $g_y(r)$ [5].

Elemental Composition (% by mass)	Medium			
	Adipose tissue	Breast	Muscle	Bone
H	11.4	10.6	10.2	3.4
C	59.8	33.2	14.3	15.5
N	0.7	3.0	3.4	4.2
O	27.8	52.7	71	43.5
Na	0.1	0.1	0.1	0.1
Mg				0.2
P		0.1	0.2	10.3
S	0.1	0.2	0.3	0.3
Cl	0.1	0.1	0.1	
K			0.4	
Ca				22.5
Density (gcm-3)	0.95	1.02	1.05	1.92

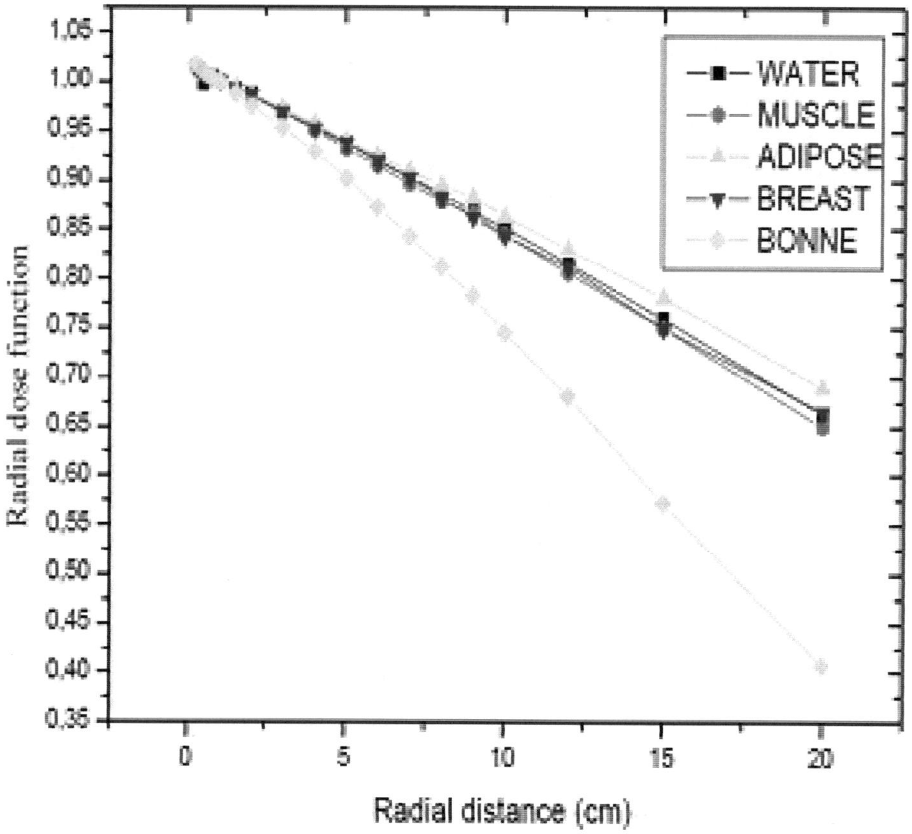

Figure 4. Comparison between $g_y(r)$ obtained in water with MCNPX and $g_y(r)$ obtained in different media with EGS5 [7].

3. Results and discussion

The results presented in the table2 refer to the calculated radial dose function $g_y(r)$ equation (2), calculated with MCNPX, which present a good consistency with the results given in other published data calculated with GEANT4, PENELOP, and EGS5 Monte Carlo codes, the little difference appears in the value of $g_y(r)$ in the radial distance near to the source, and this may depend to the difference in the geometry simulated of the source, the calculated $g_y(r)$ in water was compared to the radial dose function calculated in different medium with EGS5[6] (table2), the results (figure4) appears logical, and presents the effect of medium density for the dose deposed, $g_y(r)$ in bones decrease fast compared to the other tissues because the density of bone is the higher (1.92 gcm^{-3}) and it has a high attenuation coefficient, for the adipose tissue the density is the lowest (0.95 gcm^{-3}) that is why its $g_y(r)$ has a lower decreases. For the other media, the densities are near to the density of water (breast and Muscle) that is why their curves are superposed.

The table4 summarizes the results of the Anisotropy Function calculated with MCNPX used for this work, with the same coordinates system in figure 1.

Table 4. Anisotropy function obtained with MCNPX.

Θ(°)	Radial distance (cm)														
	0 .25	0 .5	0.75	1	2	3	4	5	6	7	8	10	12	15	20
0	–	0.9548	0.9577	0.9390	0.9443	0.9503	0.9395	0.9326	0.9297	0.9530	0.9423	0.9188	0.9517	0.9682	0.9623
8	–	0.9511	0.9781	0.9677	0.9501	0.9541	0.9486	0.9394	0.9510	0.9611	0.9420	0.9259	0.9595	0.9588	0.9638
10	–	0.9555	0.9791	0.9718	0.9552	0.9620	0.9458	0.9474	0.9569	0.9680	0.9634	0.9464	0.9639	0.9672	0.9692
15	–	0.9744	0.9837	0.9779	0.9782	0.9710	0.9807	0.9828	0.9783	0.9811	0.9724	0.9671	0.9839	0.9823	0.9781
20	–	0.9812	0.9852	0.9843	0.9781	0.9800	0.9804	0.9840	0.9768	0.9873	0.9745	0.9689	0.9877	0.9885	0.9919
25	–	0.9822	0.9896	0.9868	0.9864	0.9897	0.9812	0.9921	0.9925	0.9910	0.9883	0.9720	0.9916	0.9897	0.9928
30	–	0.9891	0.9953	0.9913	0.9898	0.9952	0.9804	0.9992	0.9924	0.9913	0.9888	0.9771	0.9951	0.9919	0.9937
40	–	0.9963	0.9992	0.9954	0.9950	0.9956	0.9844	1.0041	0.9986	0.9916	0.9897	0.9749	0.9991	0.9970	0.9991
50	–	0.9989	0.9991	0.9899	0.9876	0.9902	0.9857	0.9974	0.9999	0.9958	0.9945	0.9857	1.0001	0.9973	1.0007
60	0.9614	0.9994	0.9978	0.9997	0.9952	1.0020	0.9927	1.0015	0.9996	1.0009	0.9919	0.9855	0.9995	0.9946	0.9966
70	0.9908	1.0007	0.9991	1.0005	0.9976	1.0015	0.9940	0.9928	0.9868	0.9991	0.9970	0.9862	0.9997	0.9992	0.9996
80	0.9987	1.0003	0.9997	0.9999	0.9999	1.0105	0.9951	0.9925	0.9895	0.9970	0.9975	0.9865	1.0002	1.0005	1.0008
90	1	1	1	1	1	1	1	1	1	1	1	1	1	1	1
100	0.9968	0.9978	0.9984	0.9998	0.9950	1.0020	0.9989	0.9999	0.9911	1.0053	0.9938	0.9758	1.0027	1.0034	0.9991
110	0.9715	1.0000	0.9997	1	1.0031	1.0012	0.9982	0.9968	0.9973	0.9988	0.9990	0.9855	1.0035	1	1.0005
120	0.9614	1.0068	0.9964	0.9984	0.9975	1.0006	0.9857	0.9897	0.9995	0.9994	0.9943	0.9789	1.0019	0.9990	0.9971
130	–	0.9935	0.9924	0.9976	0.9961	0.9979	0.9856	0.9860	0.9938	0.9981	0.9902	0.9776	1.0023	0.9972	0.9997
140	–	0.9906	0.9880	0.9960	0.9974	0.9956	0.9848	0.9858	0.9918	0.9953	0.9901	0.9766	0.9932	0.9950	0.9975
150	–	0.9945	0.9865	0.9931	0.9920	0.9917	0.9812	0.9874	0.9876	0.9963	0.9862	0.9727	0.9891	0.9913	0.9970
155	–	–	0.9919	0.9880	0.9823	0.9843	0.9751	0.9793	0.9874	0.9868	0.9872	0.9666	0.9809	0.9882	0.9910
160	–	–	0.9776	0.9895	0.9802	0.9817	0.9738	0.9710	0.9863	0.9907	0.9807	0.9658	0.9810	0.9863	0.9851
165	–	–	0.9767	0.9732	0.9652	0.9706	0.9710	0.9682	0.9657	0.9760	0.9807	0.9447	0.9772	0.9744	0.9798
170	–	–	0.9351	0.9425	0.9362	0.9543	0.9355	0.9328	0.9411	0.9542	0.9710	0.9313	0.9511	0.9540	0.9567
172	–	–	0.9592	0.9350	0.9122	0.9422	0.9299	0.9147	0.9225	0.9290	0.9397	0.9087	0.9338	0.9370	0.9436
174	–	–	0.9386	0.9332	0.8982	0.9282	0.9145	0.9045	0.9027	0.9080	0.9211	0.8889	0.9185	0.9247	0.9345
177	–	–	0.9361	0.9195	0.8816	0.9260	0.9081	0.8974	0.9044	0.9005	0.9083	0.8837	0.9137	0.9200	0.9278
179	–	–	0.9451	0.9176	0.8894	0.9011	0.8984	0.8998	0.9004	0.9100	0.8903	0.8824	0.9100	0.9125	0.9224

The results of the anisotropy function (equation 3) are compared with other published data, figure 5, the function has a little difference in some values and this is because of the lower number of photon histories simulated with our workstation that limits the number in 2×10^9 photons histories. Moreover, the curve of the anisotropy function has the same form especially for the polar angle between 10 and 160 degrees, a little different for $\theta<10°$ and $\theta>160°$ can be attributed to the difference of the geometry simulated, the physic models, and also the code used in the calculations.

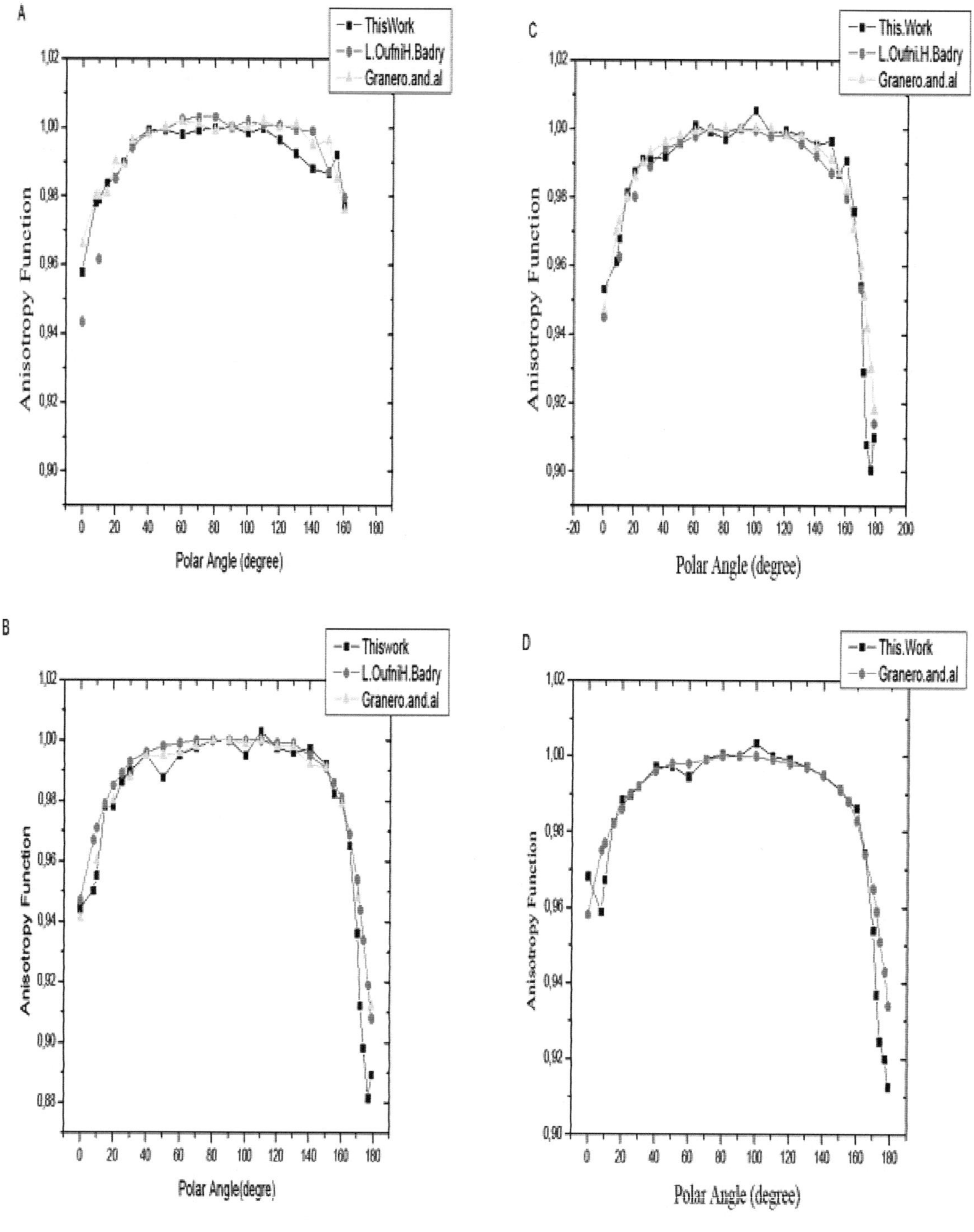

Figure 5. Values of the Anisotropy Function for different radial distances: A (r=0.75cm), B(r=2cm), C(r=7cm), D(r=15cm)

Conclusions

The curves of the radial dose function and anisotropic function obtained by MCNPX show good agreement with the published data, calculated with other Monte Carlo codes, such as Geant4, EGS5, Fluka and PENELOPE. To validate the pattern, a comparison of radial dose function was made in different medium, the results obtained are satisfying. Finally the MCNPX code can be used to investigate the dosimetric parameters for different types of sources used in brachytherapy.

REFERENCES

[1] Guerrero R Almansa J Javier F Torres A Lallena M (2014) *Phys. Med* **30** 960–967.
[2] Perez-Calatayud J Ballester F., Das R K Dewerd L A Ibbott G S Meigooni A S Ouhib Z Rivard M J Sloboda R S Williamson J F (2012) *Med. Phys* **39(5)**:2904-29.
[3] Granero D Pérez-Calatayud J Ballester F (2007) *Med. Phys.* **34** 3485–3488.
[4] Rivard MJ, Coursey BM, DeWerd LA, Hanson WF, Huq MS, Ibbott GS, et al. (2004) Med. Phys. ; **31**:633–74.
[5] ICRU (1989) Rapport no: **44**.
[6] Badry H Oufni L Ouabi H Hirayama H (2018). *Appl. Radiat. Isot.* **136** 104–110.

IOP Conf. Series: Materials Science and Engineering **783** (2020) 012020 doi:10.1088/1757-899X/783/1/012020

Design of high-performance perovskite solar cells adapted to the tandem configuration

Saida Laalioui*[1,2], Kawtar Belrhiti Alaoui [1,2], Kassem El Assali[1], Badr Ikken[2] and Abdelkader Outzourhit[1]

[1]Labaoratory for nanomaterial, energy and environment, Faculty of Sciences Semlalia, Cadi Ayyad University, PO Box: 2390, Marrakech 40000, Morocco

[2]Research Institute for Solar Energy and New Energies (IRESEN), 16, Rue Amir Sidi Mohamed Souissi - Rabat, Morocco

*Corresponding author: *laalioui@iresen.org

Abstract. Thin film based solar cells offer the added advantage of presenting a low temperature coefficient of power and are usually suggested for hot climates. Perovskite solar cells have reached today a record efficiency of 25.2%. In addition, they demonstrated a great compatibility to be used as the top cell in tandem cells, either with CIGS, CZTS, perovskite or silicon heterojunction (HIT) as the bottom cell. The purpose of this work is to design and analyse the selected planar perovskite solar cells (PSCs) reaching high power conversion efficiency and stability with layers configuration of glass substrate / FTO/ TiO_2/ $MAPbI_{(3-x)}Cl_x$/CuSCN/Au, using the SCAPS-1D software. The thickness of $MAPbI_{(3-x)}Cl_x$ and FTO was varied from 0.4 μm to 1.4 μm and from 0.02 μm to 0.5 μm respectively and the best results are observed at 1.1 μm of perovskite absorber and 0.05 FTO μm. The best performances of this device architecture, calculated with SCAPS software are cell efficiency of 27.79% with an open circuit of 1.2 V, a short circuit current of 26.04 mA/cm² and a fill factor of 88.85%.

1. Introduction

Perovskite solar cells are cells that can compete today with silicon cells. They have shown rapid growth since 2009 and they reach a record efficiency of 25.2%, which is close to the record efficiency of the best silicon heterojunction cell, which is on the order of 28.0% [1].

The perovskite materials have a wide range of properties and characteristics that are attractive for solar cells, namely: a direct gap that can easily be tuned between 1.30 and 2.5 eV, a low exciton binding energy (<10 meV), a large absorption coefficient (on the order of 1.5×10^4 cm^{-1} at 550 nm), a high carrier lifetime compared with silicon, an ambipolar conductivity and finally a good mobility of the carriers compared with the organic materials (7.5 cm^2 / Vs for the holes and 12.5 cm^2 / Vs for the electrons) [2]. Furthermore, the fabrication of this class of cells requires less sophisticated methods compared to the processes of conventional silicon cells.

A wide range of materials can be used for the different layers constituting the perovskite solar cells. This possibility offers the opportunity to get various performances that can be adapted to different applications.

The standard perovskite planar solar cell is a p-i-n type structure as showed in figure 1. The cell is composed of three main layers namely [2]: the hybrid perovskite us light absorber, an n-type semiconductor as an electron transport layer and a hole carrier layer which is a p-type semiconductor.

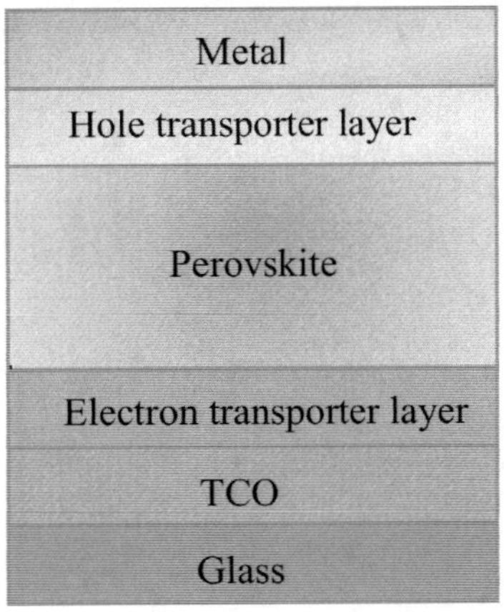

Figure 1: Standard structure of perovskite solar cell.

Perovskites refer to a family of materials with a very particular crystalline structure, of the form ABX_3. Hybrid perovskites are a subfamily of the ABX_3 perovskites and result from the combination of the organic cations A of type ($R\text{-}NH_3$), a divalent metal B of group 14, smaller than A and a halide X or combination of halides, forming a three-dimensional inorganic structure that contains the organic cation [3,4].

The current challenge is not only to improve the efficiency of the perovskite solar cell, but also to enhance its stability which depends on several factors: climatic conditions (temperature, humidity, irradiation), the nature of the perovskite absorber, the hole transport layer, the electron transport layer and the back contact [5].

The choice of the materials, the structure of the cells and the encapsulant is good way to enhance the stability of the perovskite device. So far, the best single junction solar cell containing mixed cations and halides, reached an efficiency of 21,1% and showed a good stability of up to 1000 hours at 85°C and 85% relative humidity [6].

In this paper, we report the results of a simulation work carried out by SCAPS Software, of a of perovskite solar cell with planar structure.

2. Device modelling

2.1. SCAPS program

SCAPS (Solar Cell Capacitance Simulator) is a simulation program with seven semiconductor input layers developed by a group of solar cell researcher at the department of Electronics and Information System, University of Gent, Belgium. SCAPS uses the basic semiconductor equations such as Poisson equation, continuity equation of electrons and holes.

SCAPS calculates various profiles of the defined architectures such as steady state band diagram, carrier transport, and recombination and bulk defects.

The working procedure of the SCAPS program is shown schematically in the figure 2 below:

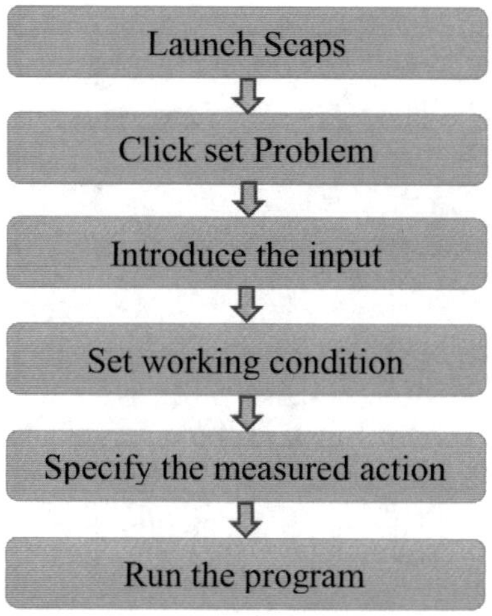

Figure 2: SCAPS working steps.

2.2. Device architecture

The choice of materials and structure is a crucial step in the design and optimization of PSCs and it is primordial to achieving high performances and approved stability. The device architecture is presented in the figure 3: First, we started with planar standard structure, with more stable and performant layers. The selection of the materials used in the simulation studies are based on the performances of these materials as shown in the literature.

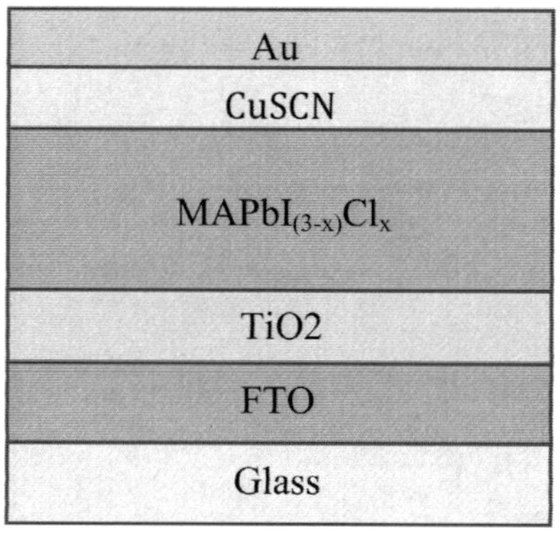

Figure 3: Device architecture of the simulated PSCs.

Indeed, the model of the device was built on transparent fluorine-doped tin-oxide (FTO) coated glass substrates, followed by a TiO_2 compact layer is a well-known inorganic material that is used as an us (Electron Transporter layer ETL) in PSCs. This choice is based on the stat-of-the-art and the previous experiences that demonstrated efficiencies up to 22.1% of the PSCs [6, 7,8, 4]

The mixed-halide perovskite $MAPbI_{(3-x)}Cl_x$ resulting from the substitution of Iodine in the well-known $MAPbI_3$ hybrid perovskite by the chlorine was used as the absorbing layer for several reasons. First, Lead-based perovskite are more stable than those based on tin, due to the instability of the later [2]. In addition, the substitution improves the proprieties of the hybrid perovskite absorber. Indeed, diffusion lengths for holes and for electrons increase from 100 nm for $MAPbI_3$ to more than 1μm in $MAPbI_{3-x}Cl_x$ [4,2]. In addition, a charge carrier mobility of up-to 33 $cm^2V^{-1}S^{-1}$ [9] was obtained in vapor-deposited $MAPbI_{3-x}Cl_x$ perovskite with a high gap of 1.55 eV making this material suitable as an absorber for the top cell in silicon-based the tandem solar cell configuration.

Copper thiocyanate (CuSCN) is one of the inorganic p-type semiconductors that is considered to be a good alternative HTM due to its high conductivity (10^{-2}–10^{-3} S cm^{-1}), stability, and transmittance; its wide band-gap (3.6 eV) and its work function of approximately −5.3 eV [10]. The first remarkable work on CuSCN HTM in PSCs was reported in 2014 by Grätzel et al., where a PSC with a 12.4% efficiency was presented, compared to that of HTM-free cell displaying only 9.5% efficiency. This was mostly attributed to a significant increase in the short-circuit current and open-circuit voltage. This improvement was due to the efficient charge extraction and collection from the excited perovskite ($MAPbI_3$ in this case) to TiO_2 and CuSCN and to the corresponding electrodes [11].

2.3. Device simulation parameters

The main parameters that must be defined to perform the simulation with SCAPS program are: band gap (Eg), electron affinity (χ), dielectric constant (ε), valence and conduction band density of states (NV and NC), charge carriers mobility (μn and μp), acceptor and donor dopant concentration (NA, ND) and thermal velocity of charge carriers (V_{thn} and V_{thp}).

The simulation parameters for each layer in perovskite solar cell are selected from previous research [12-15]. Table 1 summarizes the device parameters used in this study.

Table 1: Simulation parameters of the perovskite solar cell.

Parameters	**FTO**	**TiO_2**	**CuSCN**	**$MAPbI_{3-x}Cl_x$**
Band gap Eg (eV)	3.5	3.2	3.6	1.55
Electron affinity (eV)	4	4.2	1.7	3.9
Dielectric permittivity	9	9	10	6.5
CB effective density of states 1/cm^3	2.2 x 10^{17}	2.2 x 10^{18}	2.2 x 10^{19}	2.2 x 10^{18}
VB effective density of states 1/cm^3	2.2 x 10^{16}	1.8 x 10^{19}	1.8 x 10^{18}	1.8 x 10^{19}
Electron thermal velocity (cm/s)	1 x 10^7	1 x 10^7	1 x 10^7	1 x 10^7
Hole thermal velocity (cm/s)	1 x 10^7	1 x 10^7	1 x 10^7	1 x 10^7
Electron mobility (cm^2/Vs	20	20	100	2
Hole mobility (cm^2/Vs)	10	10	25	2
Shallow uniform donor density ND (1/cm^3)	1 x 10^{15}	1 x 10^{16}	0	1 x 10^{13}
Shallow uniform acceptor density NA (1/cm^3)	0	0	1 x 10^{18}	1 x 10^{13}

3. Results and discussion

3.1. Effect of the thickness of the perovskite

The thickness of the perovskite absorber is one of the parameters that affect significantly the performances of the PSC. The effect of thickness of the absorption layer on the solar cell performance is shown in figure 4. Indeed, when the thickness of the hybrid perovskite increases from 0.4 μm to 1.4 μm, the efficiency improves gradually and reaches its maximum when the perovskite thickness is 1.1 μm. Beyond this value, the performance of the cell decreases, which can be explained by the recombination of the photogenerated carriers before reaching the collecting layers.

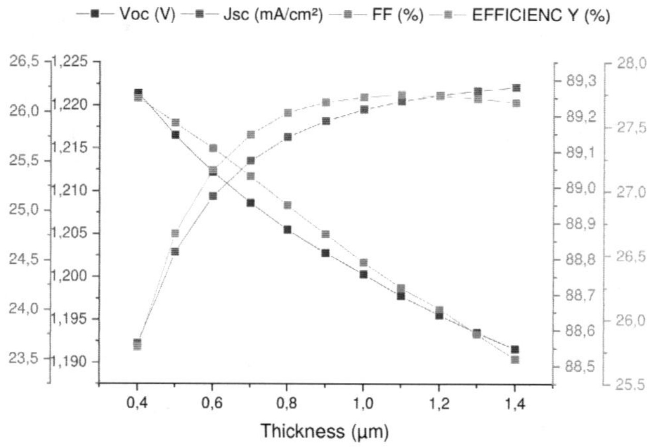

Figure 4: The effect of the perovskite thickness on the performance of the solar cell.

3.2. *Effect of the thickness of FTO*

The thickness of the FTO layer was also varied from 0.02 µm to 0.5 µm. As, shown in the figure 5 and 6 the best result was attained with 0.05 µm. However higher values of the FTO thickness lead to a significant decrease in the efficiency and the short-circuit current density, which can be explained by the parasitic absorption of the FTO layer causing a current loss of the cell.

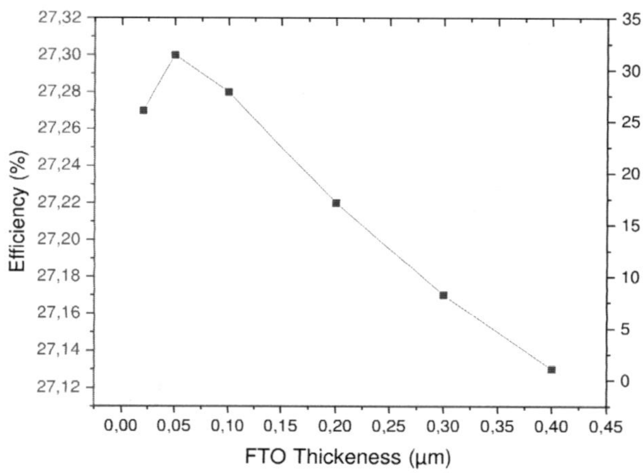

Figure 5: Variation of the cell efficiency with the variation of FTO thickness.

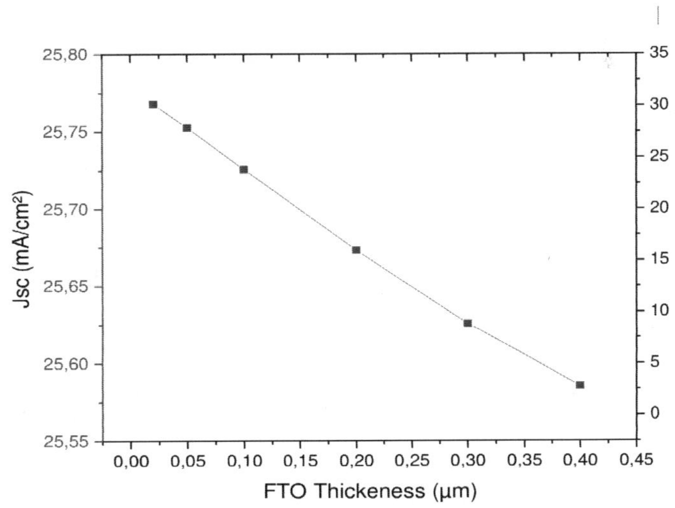

Figure 6: variation of the current density with the variation of FTO thickness.

3.3. *Effect of HTM*

The effect of the HTM was also investigated by changing the HTM and keeping the other parameter of the devices are the same. The first device was with the most used HTM; spiro-Ometad and the second with inorganic CuSCN material. The results, shown in figure 7 below, demonstrate that the use of CuSCN allows higher performances (efficiency, current density, open circuit voltage, fill factor). This is attributed to the specific properties of this material.

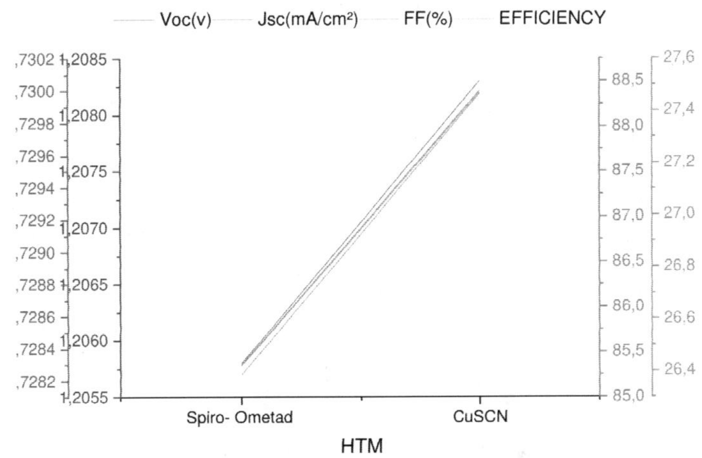

Figure 7: variation of the HTM on the performance of the solar cell.

3.4. *The optimized PCS*

Based on the previous simulation results, an optimized perovskite solar cell was obtained with the configuration: glass substrate / FTO/ TiO_2/ $MAPbI_{(3-x)}Cl_x$/CuSCN/Au with perovskite thickness of 1.1 µm and 0.05 µm for the FTO layer. As shown in the figure 8, we obtained for this optimized configuration an efficiency of 27.79% is, short circuit current density (J_{sc}) 26.12 mA/cm^2, open circuit voltage (V_{oc}) 1.19V, fill factor (FF) 88.79%.

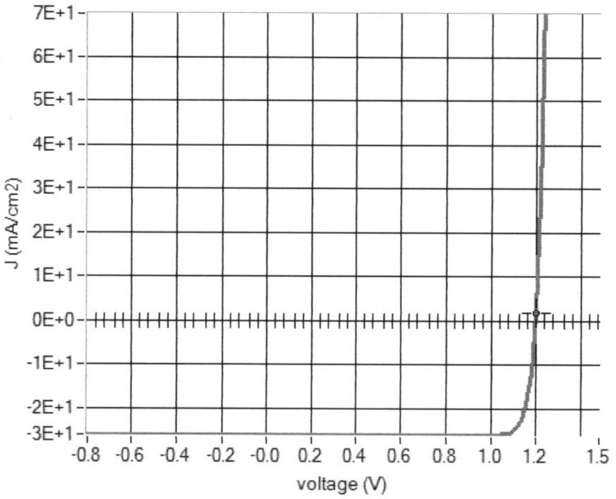

Voc (V)	Jsc (mA/cm2)	FF (%)	eta (%)
1.1979	26.125100	88.79	27.79

Figure 8: Simulated J-V of the optimized PSC.

4. Conclusions

The selection of materials plays a primordial role for the design of the perovskite solar cell. The thickness of these layers was investigated to enhance the performances of the perovskite solar cells. The best results were obtained for a perovskite absorber with double halide having a band gap of 1.55 eV and 3.9 eV electron affinity. In addition, the use of an inorganic ETL will renders the final devices more stable. The SCAPS-1D simulations resulted in the optimized perovskite solar cell with the configuration glass substrate / FTO/ TiO_2/ $MAPbI_{(3-x)}Cl_x$/CuSCN/Au and with an efficiency of 27.79%. This finding may provide the starting point for an experimental investigation of the cell.

References

[1] N. R. E. L, "Best research-cell Efficiency", 2019. [Online]. Available: https://www.nrel.gov/pv/assets/images/efficiency-chart.png.
[2] Chen Q, De Marco N, Michael Y, Song T, Chen C, Zhao H, Hong Z, Zhou H and Yang Y, 2015 *Nano Today*, **10**, 355
[3] L. J. CNRS, "Photovoltaïque : la fièvre pérovskite", 2015 [Online]. Available: https://lejournal.cnrs.fr/articles/photovoltaique-la-fievre-perovskite.
[4] Nazeeruddin M K, and Snaith H, 2017 *Nano Today*, **40**, 641–645
[5] Wang D, Wright M, Elumalai N K and Uddin A, 2016 *Sol. Energy Mater. Sol. Cells*, **147**, 255–275
[6] Yang W S, Park B W, Jung E H, Jeon N J, Kim Y C, Lee D U, Shin S S, Seo J, Kim E K, Noh J H and Seok S I, 2017 *Science*, **356**, 1376–1379
[7] Zhao Q, Li G R, Song J, Zhao Y, Qiang Y and Gao X P, 2016, *Sci. Reports*, **6**
[8] Zhao Y and Zhu K, 2015, Chem. *Soc. Rev*, **45**, 655-689
[9] Ono L K, Leyden M R and Wang S, 2015, *J. Mater. Chem. A*, **4**, 6693-6713
[10] Yang X, Wang H, Cai B, Yu Z and Sun L, 2018, *J. Energy Chem.* **27**, 650
[11] Vivo P, Salunke J K and Priimagi A, 2017, *Materials*,**10**(9), 1087
[12] Mandadapu, Vedanayakam S V and Thyagarajan K, 2017, *INDJSRT*, **10**(11).
[13] Bi D, Tress W, Dar M, Gao P, Luo J, Renevier C, Schenk K, Abate A, Giordano F, Correa Baena J P, Decoppet J D, Zakeeruddin S M, Nazeeruddin M K, Grätzel M and Hagfeldt A, 2016, *Sci. Adv*, **2**
[14] Hossain M I, Alharbi F and Tabet N, 2015, *Solar Energy*, **120**, 370-380

[15] Li F and Liu M, 2017, *J. Mater. Chem. A*, **5**, 15447-15459

Thermal Lattice Boltzmann Method for micro-Poiseuille gas flow

Youssef Elguennouni[1,*], Mohamed Hssikou[2], Jamal Baliti[3] and Mohammed Alaoui[1]

[1] University of Moulay Ismail, Faculty of Sciences, Meknes, Morocco
[2] University of Ibn Zohr, Faculty of Sciences, Agadir, Morocco
[3] University of Sultan Moulay Slimane, Polydisciplinary Faculty, Beni Mellal, Morocco

*Corresponding author e-mail: y.elguennouni@edu.umi.ac.ma

Abstract. In this paper, the thermal Lattice Boltzmann Method (TLBM) is used to simulate a gas microflow. The simulated gas is confined within an inlet/outlet microchannel and driven by a constant inlet velocity U_{in}. The bottom wall temperature is taken to be equal zero while the top wall temperature is decreasing from 1 to 0. For consistent results velocity slip and temperature jump boundary conditions at the walls are used. The rarefaction effects given by the Knudsen number, on the velocity and temperature profile are investigated. The results obtained from the TLBM are compared with those obtained using the Finite Difference Method (FDM). In addition to the good agreement observed between the two methods, the results show an interesting sensitivity of velocity and temperature profiles with the gas rarefaction degree.

1. Introduction

During the past few years, the Micro/Nanoelectro-mechanical systems (MEMS/NEMS) [1] technology has known a significant development. Many researchers have investigated the applicability of LBM in these micro and Nano-devices which are used in various fields, such as medicine, electronic, and environment control, etc.

The validity of such approach for describing a gas behavior is mainly related to the flow rarefaction degree according to the Knudsen number value. The gas flow regimes can be classified as follows:

- For $Kn < 0.001$, the continuum regime;
- For $0.001 < Kn \lesssim 0.1$, the slip regime;
- For $0.1 < Kn \lesssim 10$, the transition regime;
- For $Kn > 10$, the gas can be considered in the free molecular regime.

In the literature, very few numerical studies traits the rarefied gas microflows using LBM.

Nie et al. [2] applied the LBM method in the transitional regime for compressible flow in microchannels and micro-cavities. In their study, using the bounce-back boundary condition, they have demonstrated that the LBM can capture the fundamental behavior of flows in microchannel flows, including velocity slip, nonlinear pressure drop along the channel and mass flow rate variation with Knudsen number.

Tang et al. [3] defined a reflection coefficient r_f, which $r_f = 1$ corresponds to the pure bounce reflection, while, $r_f = 0$ is for a pure specular reflection and it can be taken between 0 and 1 to combine both of boundary conditions.

Zou and He [4] proposed a new method to specify boundary conditions in the consistency with the wall boundary condition, based on the idea of bounce-back of the non-equilibrium distribution function. When these conditions are used together with the incompressible LBGK model, the simulation results recover the analytical solution of the planer Poiseuille flow driven by a pressure gradient. The half-way wall bounce back boundary condition is also used with the pressure density inlet/outlet conditions proposed in this paper to study 2D Poiseuille flow and 3D square duct flow. The numerical results accuracy is approximately of second-order.
Tang et al. [5] proposed a thermal boundary condition for a double-population thermal lattice Boltzmann equation (TLBE) in order to solve two-dimensional Poiseuille and Couette flow. The results agree well with (FVM) and analytical solutions.
Gokaltun and Dulikravich [6] have simulated thermal micro-Couette and thermal micro-Poiseuille by using inlet/outlet boundary conditions to generate a forced convection problem. The calculation of equilibrium distributions at the wall surfaces is modified to incorporate the velocity slip and temperature jump conditions. The effect of the Knudsen number on the velocity and temperature profile is investigated.
Yasuoka et al. [7] applied a mixture of the diffuse scattering and constant wall temperature conditions to obtain thermal jump at the walls. The proposed set of schemes is validated by reference data for Fourier flow and the one around a square cylinder confined in a microchannel.
Zhang and Sun [8] have used the Chapman-Enskog expansion, to prove the consistency of the LBM method with macroscopic conservation laws. They simulated Rayleigh-Benard heat convection problem. The results are interesting and reasonable and meet the experimental data well. The stability of this scheme is also proved through different cases with a large range of Rayleigh number.
Perumal et al. [9,10] have used diffuse scattering boundary conditions and a combination of bounce-back and specular reflection to capture the slip velocity at the walls in the cases of micro-Couette, micro lid-driven cavity, and micro-Poiseuille flows.
Recently, Zarita and Hachemi [11, 12] have simulated fluid flow and heat transfer inside microchannel, slip velocity and temperature jump boundary conditions are used for the microchannel simulations with Knudsen number values covering the slip regime.
In the present study, the TLBM is used to simulate micro-Poiseuille flow for both cases of isothermal and non-isothermal channel walls temperature. The variation of slip velocity and temperature jump obtained with the current TLBM is compared with the prediction of FDM approach in the slip regime.

2. Problem statement

Two cases of micro-Poiseuille flow are treated, in the first case the temperature of the upper and lower plates is T_C, however, the temperature of the upper plate is decreasing from T_H to T_C as a function of x-coordinate in the second case as follows:

$$T(x) = T_H - \frac{(T_H - T_C)\times x}{L}. \quad (1)$$

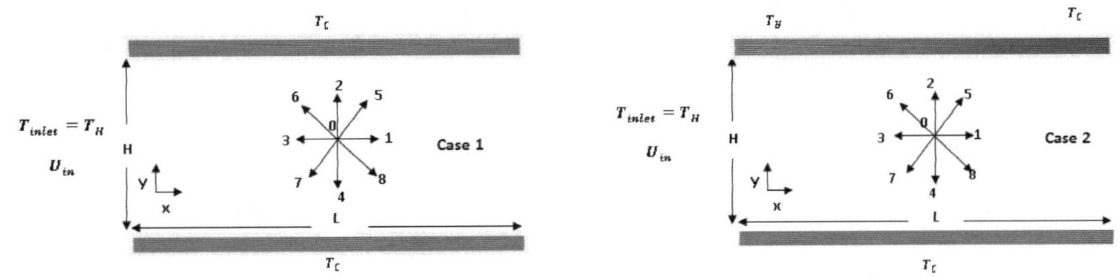

Figure 1. Domain configuration.

3. Numerical method

3.1. Thermal Lattice Boltzmann Method

A square grid and D2Q9 model is used for both distribution functions of density f and temperature g. The governing equations for these distribution functions are written according to the Bhatnagar-Gross-Krook (BGK) [13] model as:

$$f_k(x + c_k\Delta t, t + \Delta t) = f_k(x,t)\left(1 - \frac{1}{\tau_f}\right) + \frac{1}{\tau_f} f_k^{eq}(x,t), \tag{2a}$$

$$g_k(x + c_k\Delta t, t + \Delta t) = g_k(x,t)\left(1 - \frac{1}{\tau_g}\right) + \frac{1}{\tau_g} g_k^{eq}(x,t). \tag{2b}$$

In which τ_f and τ_g represent the relaxation times of density and internal energy functions, respectively. The corresponding kinematic viscosity ν and thermal diffusivity α are respectively:

$$\nu = \left(\tau_f - 1/2\right)c_s^2\Delta t,\ \ \alpha = \left(\tau_g - 1/2\right)c_s^2\Delta t.$$

In lattice: $c_s = c/\sqrt{3} = \sqrt{RT}$.

Where c_s the velocity of sound, R is is the gas constant and T is the gas temperature.

The coefficients of viscosity ν and thermal diffusion α are related to the Prandtl number Pr by $Pr = \nu/\alpha$.

At the equilibrium state, the Maxwell distribution functions f_k^{eq} and g_k^{eq} are written in the Taylor expansion as:

$$f_k^{eq}(k,i,j) = w_k\rho(i,j)\left[1 + 3\frac{\mathbf{c_k.u}}{c^2} + \frac{9}{2}\frac{(\mathbf{c_k.u})^2}{c^4} - \frac{3}{2}\frac{\mathbf{u}^2}{c^2}\right], \tag{3a}$$

$$g_k^{eq}(k,i,j) = w_kT(i,j)\left[1 + 3\frac{\mathbf{c_k.u}}{c^2} + \frac{9}{2}\frac{(\mathbf{c_k.u})^2}{c^4} - \frac{3}{2}\frac{\mathbf{u}^2}{c^2}\right]. \tag{3b}$$

Where the weight factors w_k are $w_0 = 4/9$, $w_{1-4} = 1/9$, and $w_{5-8} = 1/36$.

The discrete scheme D2Q9 is characterized by:

$$c_k = \begin{cases} (0,0), & k = 0 \\ c\left(cos\left[(k-1)\frac{\pi}{2}\right], sin\left[(k-1)\frac{\pi}{2}\right]\right), & k = 1-4 \\ c\left(cos\left[(k-5)\frac{\pi}{2} + \frac{\pi}{4}\right], sin\left[(k-5)\frac{\pi}{2} + \frac{\pi}{4}\right]\right), & k = 5-8 \end{cases} \tag{4}$$

The macroscopic density ρ, velocity $\mathbf{u}$ and internal energy by unit of mass e, are obtained by the following relations:

$$\rho = \textstyle\sum_{k=0}^{8} f_k, \tag{5a}$$

$$\rho\mathbf{u} = \textstyle\sum_{k=0}^{8} \mathbf{c_k}\, f_k, \tag{5b}$$

$$\rho e = \textstyle\sum_{k=0}^{8} g_k. \tag{5c}$$

Where $e = DRT/2$, in which D is the number of physical dimensions (equal to 2 in the current work).

3.2. Flow and temperature boundary conditions

In this study, the boundary conditions proposed by Zou and He [4] are used at the inlet and a simple extrapolation scheme is used for velocity at the outlet.

Slip boundary condition are used at the top and bottom walls [14]. At the bottom wall the unknown distributions functions (f_2, f_5, f_6) are calculated as follows:

$$u_{x,y=0} = \lambda\left(\frac{4\,u_{x,1} - u_{x,2}}{2 + 3\,\lambda}\right), \tag{6a}$$

$$\rho_{bot} = f_0 + f_1 + f_3 + 2(f_4 + f_7 + f_8), \tag{6b}$$

$$f_2 = f_4, \tag{6c}$$

$$f_5 = \frac{\rho_{bot}(1 + u_{x,y=0}) - (f_0 + f_2 + f_4)}{2} - (f_1 + f_8), \tag{6d}$$

$$f_6 = \frac{\rho_{bot}(1 - u_{x,y=0}) - (f_0 + f_2 + f_4)}{2} - (f_3 + f_7). \tag{6e}$$

Where the gas mean free path is defined as $\lambda = Kn\,H$, Kn is the Knudsen number and H is the total number of lattice nodes in the vertical direction.

Temperature jump boundary condition is written at the bottom wall as [14]:

$$T_{y=0} = \frac{C_{jump}(4\,T_1 - T_2) + 2\,T_{wall}}{2 + 3\,C_{jump}}. \tag{7}$$

C_{jump} is the temperature jump coefficient defined as:

$$C_{jump} = \phi\left(\frac{2\,\gamma}{(1+\gamma)Pr}\right)\lambda = k\,\lambda. \tag{8}$$

In which ϕ represents the thermal-accommodation coefficient and is assumed to be unity, γ is the specific heat ratio.

3.3. Nusselt number calculation

The heat transfer characteristic of the flow can be determined using the Nusselt number calculated on the bottom wall as follows:

$$Nu(x) = \frac{2\,H\left(\frac{\partial T}{\partial y}\right)_{wall}}{T_{wall}-T_{bulk}}. \quad (9)$$

3.4. Finite Difference Method

The advection-diffusion equation in 2D can be expressed by

$$T^{n+1}(i,j) = \frac{1}{\tau_g}\left(\frac{T^n(i+1,j)+T^n(i-1,j)+T^n(i,j+1)+T^n(i,j-1)}{2}\right) + T^n(i,j)\left(1-\frac{1}{\tau_g}\right) - u\Delta t\left(\frac{T^n(i,j)-T^n(i-1,j)}{\Delta x}\right). \quad (10)$$

Where n is the number of the time step Δt, and $\frac{1}{\tau_g} = \frac{2\,\alpha\Delta t}{\Delta x^2}$.

Temperature jump boundary conditions used in TLBM approach are also used in FDM approach at the bottom and the top walls and the following theoretical velocity profile across a microchannel used in the FDM method is:

$$u(y) = \frac{6\,U_{mean}\left(\frac{y}{H}-\frac{y^2}{H^2}+\frac{2-\sigma}{\sigma}Kn\right)}{\left(1+6\frac{2-\sigma}{\sigma}Kn\right)}. \quad (11)$$

Where σ is the momentum-accommodation coefficient which is assumed to be equal unity, for full diffuse reflection [15].

4. Results and discussion

In order to validate the method presented in this paper, 2D channel flow between two parallel plates at rest has been simulated. Two cases of micro-Poiseuille flow are treated, in the first case the temperature of the upper and lower plates is T_C, however, the temperature of the upper plate is decreasing from T_H to T_C in the second case. To describe easily the results it is more convenient to use the following normalization for temperature: $\theta = \frac{T-T_C}{T_H-T_C}$, the temperatures T_{inlet}, T_H, and T_C become respectively $\theta_{inlet} = \theta_H = 1$, and $\theta_C = 0$ [11]. In this study, we take $Pr = 0.71$, $k = 1.667$ and Reynolds number is fixed at $Re = 10$.

4.1. Validation test

To ensure the validation of the method TLBM, the first case is treated and the results are compared with the results presented in Zarita et al. [11] and the FDM approach (see Figs. 2, 3, and 5). To observe the rarefaction effect on the gas flow behavior, the velocity profiles for $Kn = 0.01$ and 0.08 (Fig. 2), and the temperature profiles for $Kn = 0.01$, 0.05, and 0.08 (Fig. 3) are plotted. Figures show the effect of Kn on the velocity slip and temperature jump at the centerline of the microchannel. Good agreement between the analytical solution and TLBM approach for horizontal velocity along the centerline vertical axis is observed. It is shown that velocity slip is sensitive to the rarefaction degree near the longitudinal walls. Therefore, the kinetic slip-jump phenomena are well captured by TLBM in the slip regime mainly encountered in MEMS devices. To compare the transient time of both solutions given by FDM and TLBM approaches, the temperature profile is plotted as a function of time steps number for $Kn = 0.01$, and 0.1. For a given Kn, the TLBM and FDM require the same time to reach steady state (7000 steps) and with increasing of Kn, the temperature value in the equilibrium state at the bulk increases (Fig. 4). The convergence of the Nusselt number to a constant value is reached as soon as the velocity and the temperature become constant (Fig. 5), and the temperature contours for $Kn = 0.05$ are shown in Figs. 6.

4.2. Mesh size independence

Table 1 shows the effect of the mesh on the horizontal velocity, temperature near the inlet ($x/L = 0.05$, $y/H = 0.5$), and on the average Nusselt number by using TLBM approach for the channel aspect ratio $AR = 4$. To save the computation time, the grid size 200 × 50 is enough to give good agreement data.

Table 1. Effect of the mesh on the velocity, temperature and average Nusselt number for $Kn = 0.05$.

Mesh	180 × 45	200 × 50	220 × 50
u/U_{in}	1.08	1.08	1.08
θ	0.963	0.964	0.963
Nu	5.27	5.26	5.26

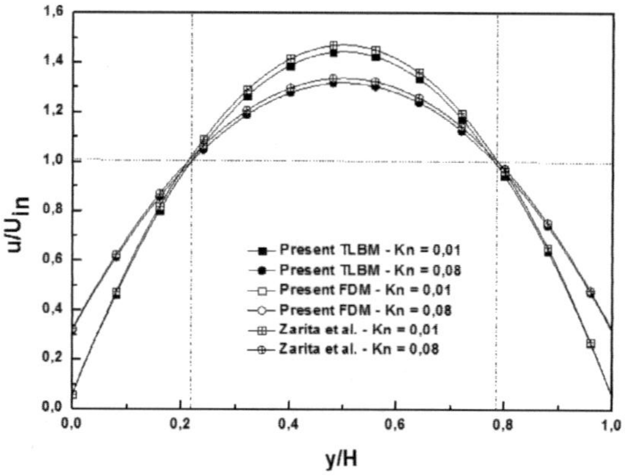

Figure 2. Velocity profiles along the centerline vertical axis according to *Kn*.

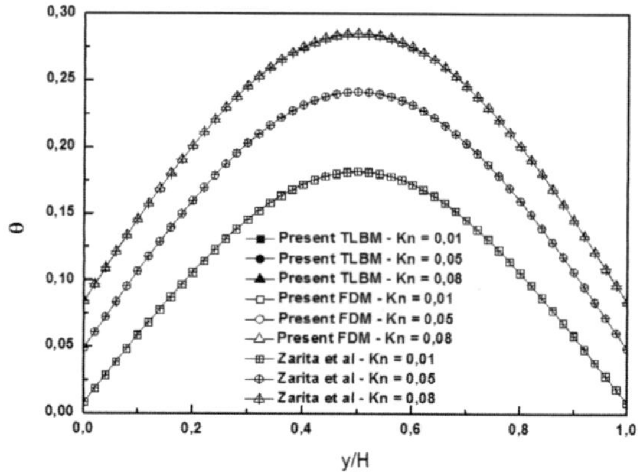

Figure 3. Temperature profiles at centerline along the vertical axis according to *Kn*.

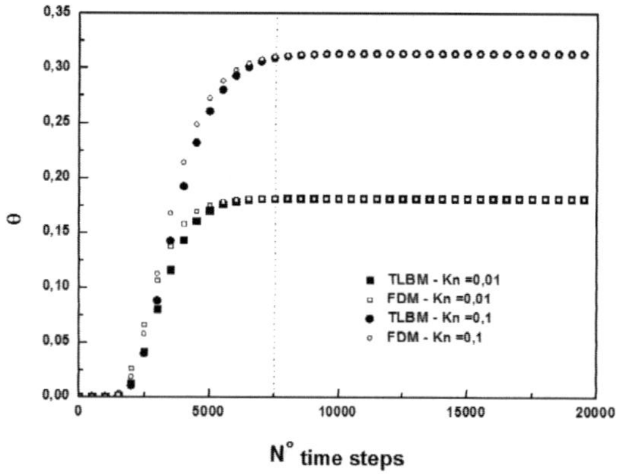

Figure 4. Evolution of the temperature at the center of the microchannel.

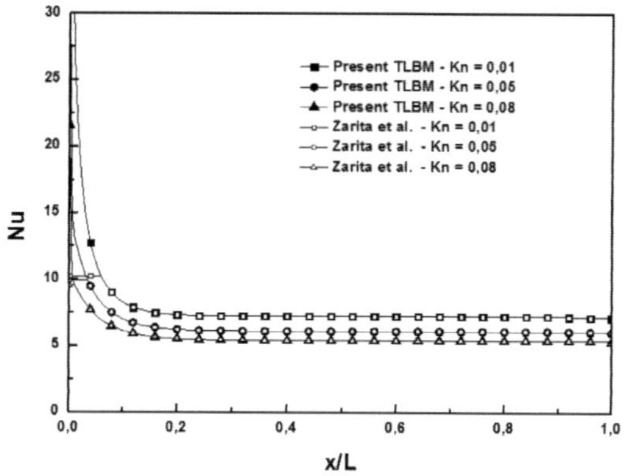

Figure 5. Nusselt number variation along the microchannel according *to Kn*.

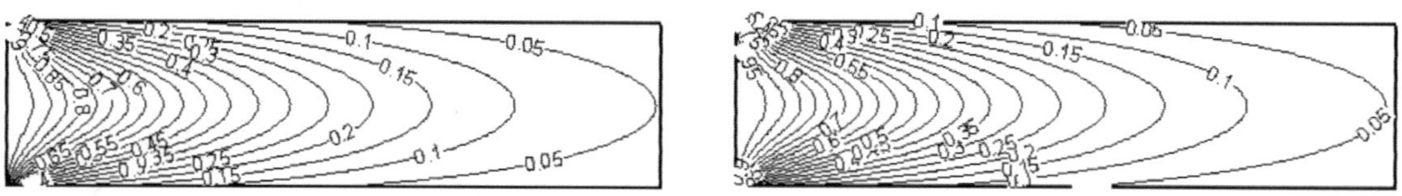

(a) TLBM - $Kn = 0.05$.

(b) FDM - $Kn = 0.05$.

Figure 6. Temperature isolines (isotherms) in the microchannel.

IOP Conf. Series: Materials Science and Engineering **783** (2020) 012021 doi:10.1088/1757-899X/783/1/012021

4.3. Numerical results and analysis

By using TLBM approach, both cases give the same velocity profiles since Eq. (5b) is independent of the temperature. Using the slip boundary condition, the so-called velocity slip phenomenon is captured. By increasing the rarefaction degree (*Kn*), the slip velocity increases (see Figs. 7a, and 7b) and the amplitude of the horizontal velocity at the center of the microchannel decreases at the steady state that reached after approximately 2200 steps (Fig. 8). Fig. 9 shows that by increasing the rarefaction degree (*Kn*), the temperature curve at the channel center reaches a high value (for $Kn = 0.1$ - $\theta = 0.556$). In the horizontal direction, the temperature decreases quickly near the bottom wall ($y/H = 0.01$ and $y/H = 0.05$) and reaches a constant value almost equal zero for $Kn = 0.01$ but this form becomes inclined and the outlet temperature becomes larger because of the gas rarefaction ($Kn = 0.1$), good agreement between the analytical solution of FDM and TLBM as shown in figs. 10a, 10b, 11a, and 11b. However, in the vertical direction, the temperature profile loses its parabolic form. The temperature is sensitive to the Knudsen number for both walls, and by increasing the rarefaction degree *Kn*, the temperature jump increases (see Figs. 11a, 11b, and Table 2). The amplitude of the temperature near the channel inlet ($x/L = 0.01$ and $x/L = 0.05$) takes a value almost equal 1 when $y/H > 0.5$. For both cases, by increasing the value of Knudsen number, the average value of the Nusselt number decreases, but in the second case, this decreasing becomes more pronounced (Table 3, and Fig. 12). The temperature contours for $Kn = 0.05$ are shown in Figs. 13a, and 13b and it is observed that both approaches give the same temperature isolines. This shows that non-uniform wall temperature affects significantly the thermal transfer in the channel.

The temperature jump is calculated at the center of the top wall as follows:

$$\theta_{jump} = \theta\left(\frac{x}{L} = 0.5\right) - \theta_{wall}. \tag{12}$$

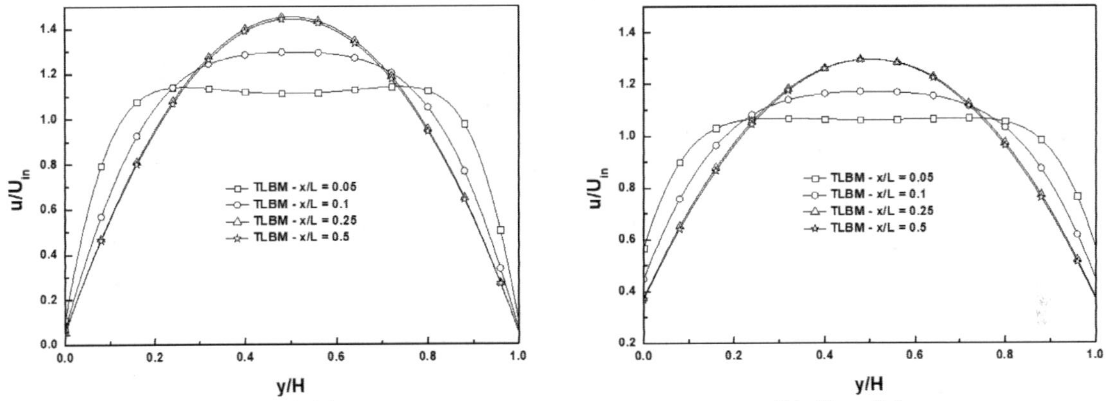

(a) $Kn = 0.01$. (b) $Kn = 0.1$.

Figure 7. Velocity profile along the vertical axis.

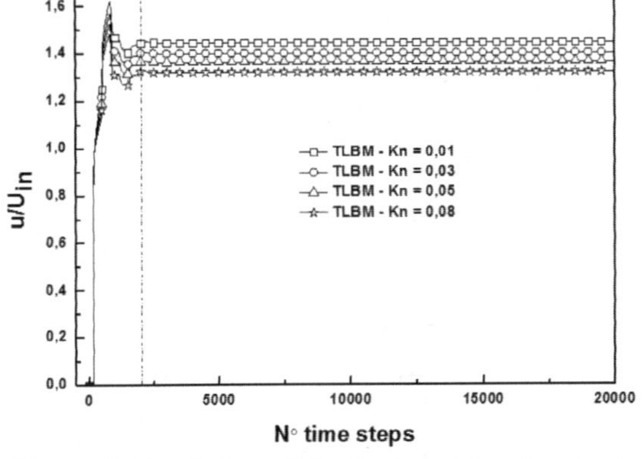

Figure 8. Evolution of the horizontal velocity at the center of the microchannel.

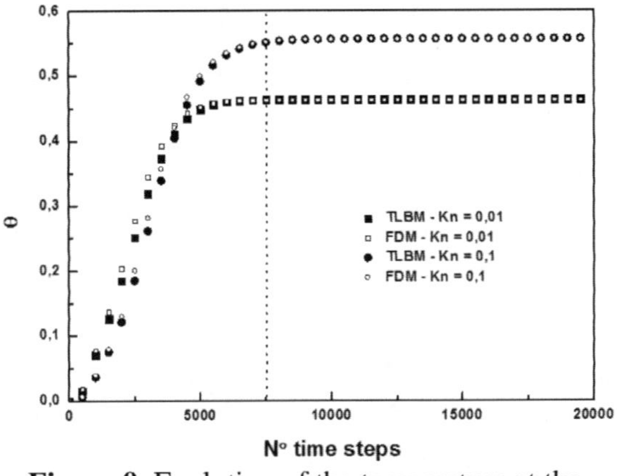

Figure 9. Evolution of the temperature at the center of the microchannel.

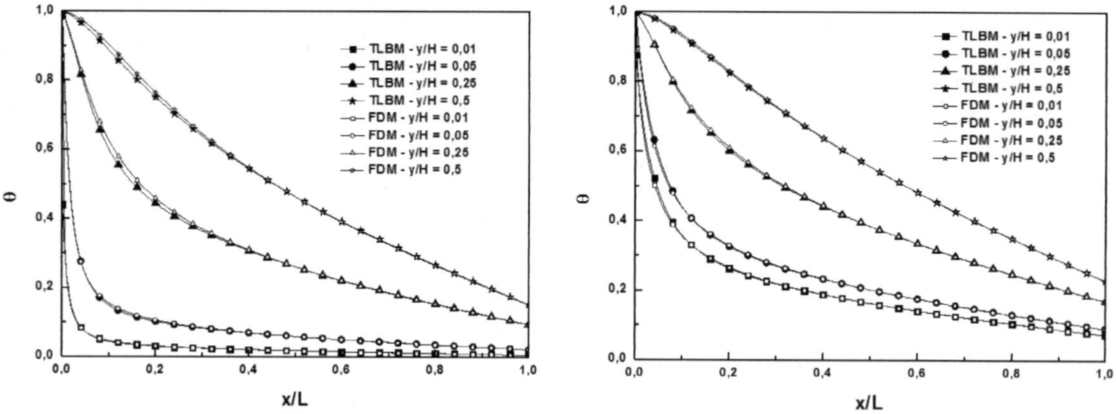

(a) $Kn = 0.01$. (b) $Kn = 0.1$.

Figure 10. Temperature profile along the horizontal axis.

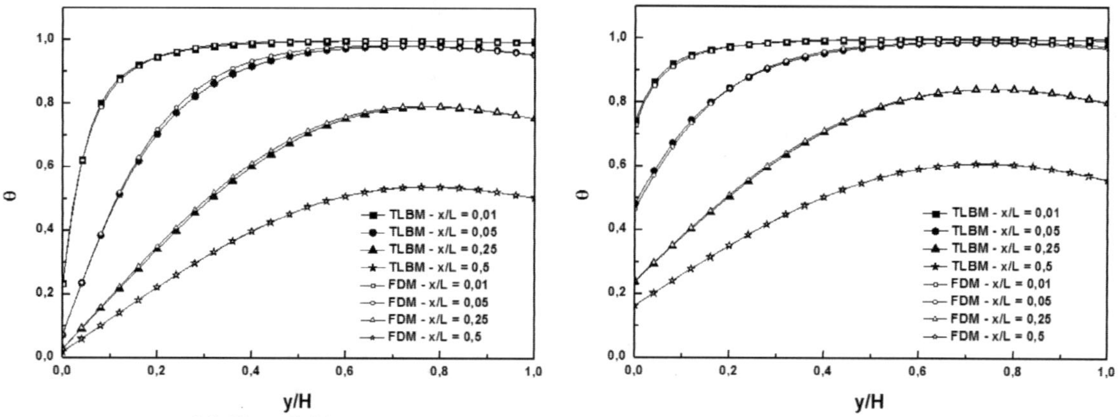

(a) $Kn = 0.01$. (b) $Kn = 0.1$.

Figure 11. Temperature profile along the vertical axis.

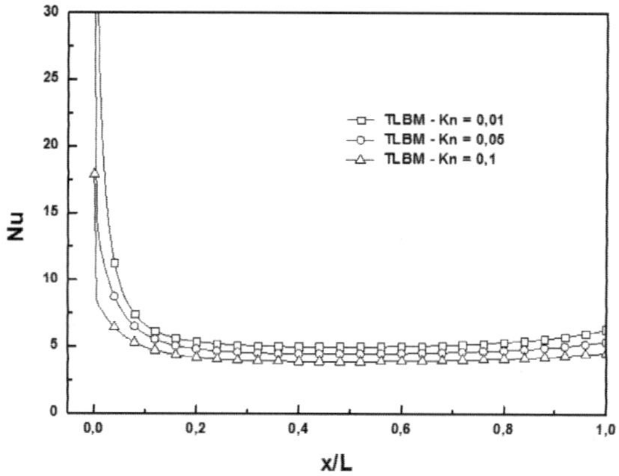

Figure 12. Nusselt number variation along the microchannel according to Kn.

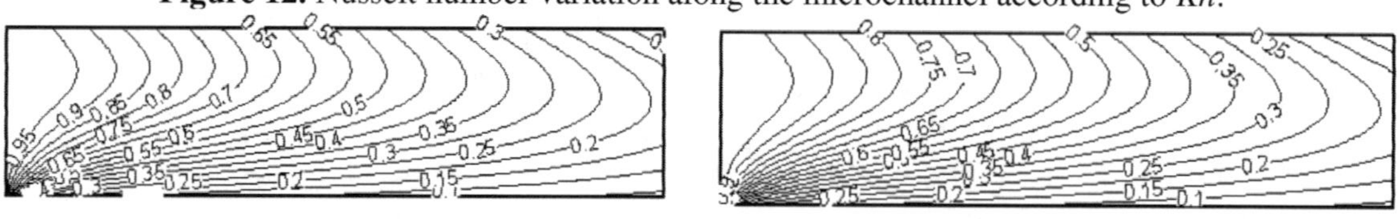

(a) TLBM - $Kn = 0.05$. (b) FDM - $Kn= 0.05$.

Figure 13. Temperature isolines (isotherms) in the microchannel.

Table 2. Effect of *Kn* on the temperature jump at the center of top wall.

Kn	**0.01**	**0.1**
TLBM	0.00401	0.05805
FDM	0.00388	0.05692

Table 3. Value of average Nusselt number at different *Kn* by TLBM approach.

	Kn	0.01	0.05	0.1
First case	*Nu*	8.34	6.62	5.31
Second case	*Nu*	6.43	5.26	4.39

5. CONCLUSION

The TLBM with slip and jump boundary conditions is used to simulate the gas flows in microchannel with isothermal and non-isothermal walls. It is observed that non-isothermal wall has a significant effect on both thermal and hydrodynamic flow behaviors. The slip-jump boundary conditions allow capturing of non-equilibrium effect near the walls. Good agreement is observed between the approaches TLBM and FDM. So, regarding its fast convergence, the TLBM is, therefore, a good alternative that can be used to describe the gas microflows usually encountered in the MEMS/NEMS devices. Unlike the pure kinetic methods, like the direct simulation of Monte Carlo (DSMC), which needs a long time to achieve satisfactory results.

References

[1] Karniadakis G Beskok A Aluru N (2005) *Springer New York*
[2] Nie X Doolen G D Chen S (2002) *J. Stat. Phys.* **107** 279
[3] Tang G H Tao W Q He Y L (2003) *Int. Conf. Microchannels Minichannels* **1** 389
[4] Zou Q and He X (1997) *Phys. Fluids* **9** 1591
[5] Tang G H Tao W Q He Y L (2005) *Phys. Rev. E* **72** 016703
[6] Gokaltun S Dulikravich G S (2014) *International Journal of Heat and Mass Transfer* **78** 796
[7] Yasuoka H Kaneda M Suga K (2016) *Phys. Rev. E* **94** 013102
[8] Zhang T Sun S (2018) *International Conference on Computational Science* 149
[9] Perumal D A Kumar G V S Dass A K (2009) *Conference Proceedings*
[10] Perumal D A Kumar G V S Dass A K (2010) *CFD Letters* **2** 25
[11] Zarita R Hachemi M (2014) *Micro and Nano Flows Conference*
[12] Zarita R Hachemi M (2019) *Frontiers in Heat and Mass Transfer* **12**
[13] Bhatnagar P L Gross E P Krook M (1954) *Phys. Rev.* **94** 511
[14] Tian Z, Zou C Liu Z, Guo Z Liu H Zheng C (2006) *International Journal of Modern Physc. C* **17** 603
[15] Drake R M Jr Eckert E G R (1972) *McGraw-Hill Newyork*

IOP Conf. Series: Materials Science and Engineering **783** (2020) 012022 doi:10.1088/1757-899X/783/1/012022

Noncovalent functionalization of carbon nanotubes by polyaniline molecules: Raman analysis

Jamal Chenouf[1], Mourad Boutahir[1], Brahim Fakrach[1, 2], Abdelhai Rahmani[1], Hassane Chadli[1], Abdelali Rahmani[1]
[1]Laboratoire d'Etude des Matériaux Avancés et Applications, Université Moulay Ismail, Faculté des Sciences Meknes, Morocco
[2]Laboratoire de Physique Théorique et Appliquée, Université Sidi Mohammed Ben Abdellah, Faculté des Sciences Dhar El Mahraz Fez, Morocco

*Corresponding author e-mail: chenouf2021jamal.gmail.com

Abstract. Polyaniline/single-walled carbon nanotube composite (PANI4@SWCNT) has been studied theoretically by means of Raman spectroscopy calculations. We evidence that the structural organization of the nanohybrid systems depends on the diameter of carbon nanotubes. To study the stability of PANI4 molecules encapsulated in the cavity of the SWCNTs, the optimum configurations of PANI4 inside carbon nanotubes are derived from minimum energy calculations using a convenient Lennard-Jones expression. We discussed also two important Raman features of SWCNTs before and after encapsulation of PANI4 including the G-band and the radial breathing mode (RBM) taking into account its metallic or semiconducting character. The G-band shift suggests a significant charge transfer between the PANI4 and the nanotubes.

1. Introduction

Among the different types of solar cells developed in recent decades, organic solar cells have interested many researchers from academia and industries due to their low technological cost, lightweight, and large-area flexible fabrication [1]. However, the most crucial challenge for organic solar cells is their low efficiency and lifetime [2,3].

Several conducting polymers have been employed as promising materials for the active layer in organic solar cells [4]. Amongst, polyaniline (PANI) is an exceptional and promising candidate for solar cell applications because of its easy synthesis route, its good processability, beneficial environmental stability, and controllable electrical properties [5]. Even more interestingly, PANI offers the opportunity to obtain low bandgap polymer, which is an effective factor to increase efficiency [6]. Nevertheless, it is necessary to improve their energy efficiency and durability for the development of sustainable technology.

Hence, in recent years, researchers have been attracted to the use of highly conductive materials in these solar cells due to avoiding carrier recombination [7]. Carbon nanotubes (CNTs) by regarding their unique structural, chemical, mechanical, thermal and electronic properties [8,9] have been integrated successfully into organic photovoltaic devices as part of the active layer and can increase the efficiency in polymeric solar cells about 3% [10].

However, to improve the efficiency of polymer/CNT based solar cells, particularly the combination of CNT with PANI for using in the organic solar cells, have interested a growing number of researchers worldwide, in order to improve the properties of these composites and to increase efficiency [11]. Indeed, the low mobility of charge carrier in conjugating polymers is one of the main limiting factors lowering the efficiency of charge separation, and, by extension, the efficiency of devices [12]. This problem is overcome by adding CNT to the polymer. Creating an interface between the conversion element and the one used for conduction, allowing then separate the charge carriers and create and collect a current [13].

Various CNT functionalization approaches that have been used to improve CNT incorporation into polymer layers in a covalent or non-covalent manner [14]. However, it is important to report that in the case of SWCNT, covalent functionalization perturbs some of the aromatic nature of the nanotubes, leaving defects in the structure of the nanotube, which significantly affect both mechanical and electrical properties [15]. To control these limitations and exploit the benefits respective of SWCNTs and photoactive molecules, an elegant solution is to encapsulate polymer into CNT to form new hybrid systems with adjustable optoelectronic properties. The insertion of oligomers inside the cavity of the carbon nanotubes is particularly attractive when the preservation of intrinsic properties of CNTs is desired, mainly electronic or charge transport behavior [16]. In addition, the encapsulation strategy is an innovative technique that can overcome the degradation problem as reported by Yanagi et al [17] when the degradation of π-conjugated molecules might be overcome by encapsulating them inside carbon nanotubes.

In this regard, we study in the present paper the PANI4 encapsulated in SWNTs (PANI4@SWCNT). The total energy minimization study of PANI4 @ SWCNT to demonstrate the most stable structure of the hybrid system with an optimal diameter is reported using a convenient Lennard-Jones expression of the van der Waals intermolecular potential. We investigate also the PANI4-tube interactions effect on the vibrations of PANI4@SWCNT hybrid system using the non-resonant Raman spectra. The calculations have been performed in the framework of the bond-polarization theory using the spectral moment's method (SMM) [18] and have brought information concerning the interaction between the PANI4 molecules and carbon nanotubes by analysis of both RBM and G-band modifications.

2. Model, Method & Approach

The configurations of PANI4 oligomers inside the carbon nanotubes are derived from minimization energy calculations performed as follow. We encapsulate PANI4 inside carbon nanotubes. The inserted molecule position is centralized along the main axis of the tube, and the diameter of the tube can vary during our structural relaxation procedure. The dynamical matrix of oligomers encapsulated in SWCNT is built block by block from the dynamical matrices of each subsystem: the dynamical matrix of the oligomer and SWCNT forms the bidiagonal block of the dynamical matrix of the system, while the dynamical matrix associated with the SWCNT oligomers interactions form the off-diagonal block. The dynamical matrix of free oligomer molecules is calculated using DFT as implemented inside the Spanish initiative for Electronic Simulations with Thousands of Atoms (SIESTA) package [19]. Core electrons were replaced by nonlocal norm-conserving pseudopotentials. The valence electrons were described by a double singly polarized basis set. The localization of the basis is controlled by an energy shift of 50 meV. Real space integration was performed on a regular grid corresponding to a plane-wave cut around 360 Ry. The first Brillouin zone was sampled according to the Monkhorst- Pack scheme by using 10 k-points along the molecule direction. We considered a vacuum size of 12Å to avoid interactions between adjacent systems.

For the SWCNT, the C-C interactions are computed using the force constant model introduced by Saito et al. [20]. This model was previously used by our group to study the non-resonant Raman spectra of single-walled carbon nanotubes (SWCNTs) and double-walled carbon nanotubes (DWCNTs) as a function of their lengths, diameters and chiralities [21-24]. The interactions between oligomer-SWCNT are described according to the van der Waals potential. We choose a Lennard Jones type:

$$U_{LJ}(r) = 4\varepsilon\,[(\sigma/r)^{12} - (\sigma/r)^{6}] \tag{1}$$

with parameters: $\varepsilon = 2.964$ meV and $\sigma = 0.3407$ nm for Carbon atoms. $\varepsilon = 7.281$ meV and $\sigma = 0.325$ nm for Nitrogen atoms. $\varepsilon = 0.67$ meV and $\sigma = 0.1069$ nm for Hydrogen atoms. The Nitrogen -Carbon and Carbon-Hydrogen parameters were calculated using the rule of mixtures of Lorentz-Berthelot:

$$\varepsilon_{AB}=(\varepsilon_A\varepsilon_B)^{1/2} \quad \text{and} \quad \sigma_{AB}=(\sigma_A + \sigma_B)/2 \tag{2}$$

3. Results and discussions

3.1. Structural study of PANI4 encapsulated in SWCNTs

In order to study the structural organization of the nanohybrid systems to determine the optimal diameter for which PANI4@SWCNT is more stable, we performed minimization energy calculations on PANI4@SWCNT using Lennard-Jones potential. The optimized configuration corresponds to an isolated chain inserted in the carbon nanotube and centered along the nanotube main axis. Figure 1 displays the results obtained of carbon nanotubes with different diameters filled with PANI4 and gives clear evidence that insertion PANI4 is favorable for tubes with diameter around 1.25nm. Therefore, the most configurations energetically favorable for the insertion of a single molecule aligned along the axis of the carbon nanotube and located at its center, are those which correspond to this optimal diameter.

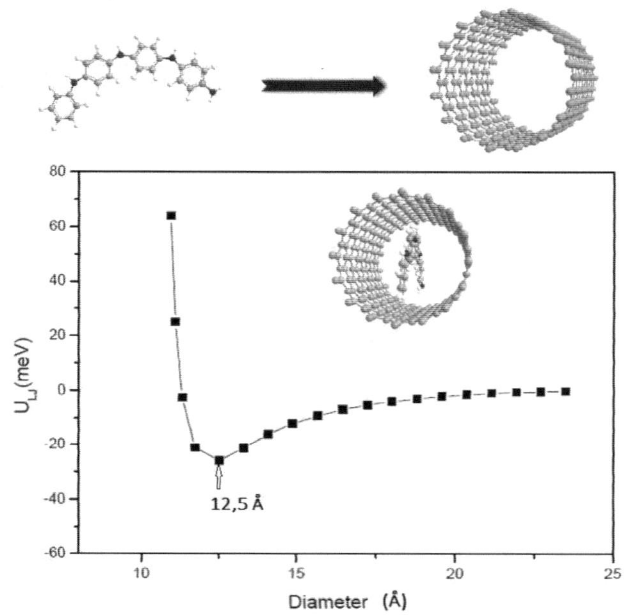

Figure 1. Minimization energy calculations on PANI4@SWCNT and schematic of the encapsulation of PANI4 molecules in SWNT. The blue color indicates the Nitrogen, gray the Carbon, and white the hydrogen atoms.alculated $I_Z(\omega)$ (dashed lines) and $I_X(\omega)$ (solid lines) of (10,10)@(15,15) CBN nanotube with their corresponding isolated nanotubes.

3.2. Raman spectra of PANI4 confined in SWCNTs

Raman spectroscopy plays an important role in the study and characterization of graphitic materials and widely used to characterize fullerenes, graphene and carbon nanotubes [25,26]. For carbon nanotubes, Raman spectroscopy can give much information about nanotube diameter, chirality, curvature, the optical energy gap, the metallic or semiconducting behavior, and more. In this article, two important Raman features of SWCNTs are discussed before and after encapsulation of PANI4 including the G-band, and the radial breathing mode (RBM). As we can see in the previous study of the stability of the nanohybrid systems, the optimal diameter calculated for which PANI4@SWCNT is more stable was around 1.25nm. In order to analysis the Raman spectra of PANI4@SWCNT, we consider six nanotube chiralities (n,m) with diameters between 1.20 and 1.29nm taking into account the metallic or semiconductive nature of SWCNTs.

3.2.1. Hybrid systems with metallic nanotubes

For metallic SWCNTs, we consider three nanotube chiralities (n,m) with diameters around 1.25nm such as: (9,9) with d_t=1.22nm, (11,8) with d_t=1.29 nm, and (13,4) with d_t=1.20nm. The ZZ-polarized Raman spectra of metallic SWCNTs and hybrid systems after minimizations are presented in figure 2, focusing mainly on the radial breathing mode (RBM) frequency region and G-band one.

After the encapsulation, the main RBM peak was down-shifted from 185 to 184 cm^{-1}, from 173 to 172 cm^{-1}, and from 183 to 181cm^{-1} for the selected hybrid systems, PANI4@(13,4), PANI4@(11,8), and PANI4@(9,9) respectively as shown in figure 2(left). The origin of this lower frequency shift of the RBM can be attributed to the dispersive interactions (van der Waals) [27-29].

Concerning the high-frequency region, where the tangential mode (G-band) is observed. The G-band in PANI4@(13,4) nanotube shifted up by about 3 cm^{-1} as shown in figure 2(right). These shifts to higher frequencies in the G-band indicate a charge transfer from the SWCNT to the molecule. For PANI4@(11,8) and PANI4@(9,9) hybrids, no shift of G-band could be observed suggesting that charge transfer is not occurring.

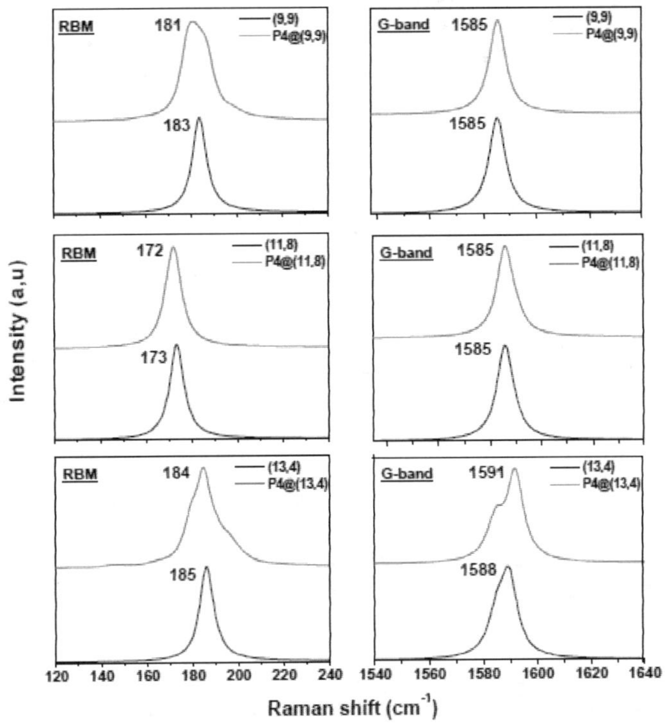

Figure 2. Raman spectra of SWCNTs and PANI4@SWCNTs in the RBM range and G-band.

3.2.2. Hybrid systems with semiconducting nanotubes

Similarly, for semiconducting SWCNTs, we consider three nanotube chiralities (n,m) with diameters around 1.25nm such as: (15,2) with d_t=1.25nm, (14,4) with d_t=1.28 nm, and (10,9) with d_t=1.28nm. The ZZ-polarized Raman spectra of semiconducting SWCNTs and hybrid systems after minimizations are presented in figure 3, focusing mainly on the radial breathing mode (RBM) frequency region and G-band one.

Firstly, in the low-frequency region, where the radial breathing modes (RBM) are observed, a lower frequency shift of the RBM can be attributed to the van der Waals dispersion interactions. The RBM clearly exhibits a frequency downshift after encapsulation by about 1cm^{-1} for both PANI4@(15,2) and PANI4@(14,4) nanotubes, and 4cm^{-1} for PANI4@(10,9) nanotube as shown in figure 3(left).

Concerning the high-frequency region, where the tangential mode (G-band) is observed, as one can see in figure 3(right), the G-band in SWCNTs shifted up by about 3cm^{-1} for both PANI4@(15,2) and PANI@(14,4) hybrids. These shifts to higher frequencies in the G-band indicate a charge transfer from the SWCNT to the PANI4 molecule. In the case of PANI4@(10,9) hybrid, G-band shifted down by about 7cm^{-1}. This downshift of G-band can be explained by a negative charge transfer on the nanotubes.

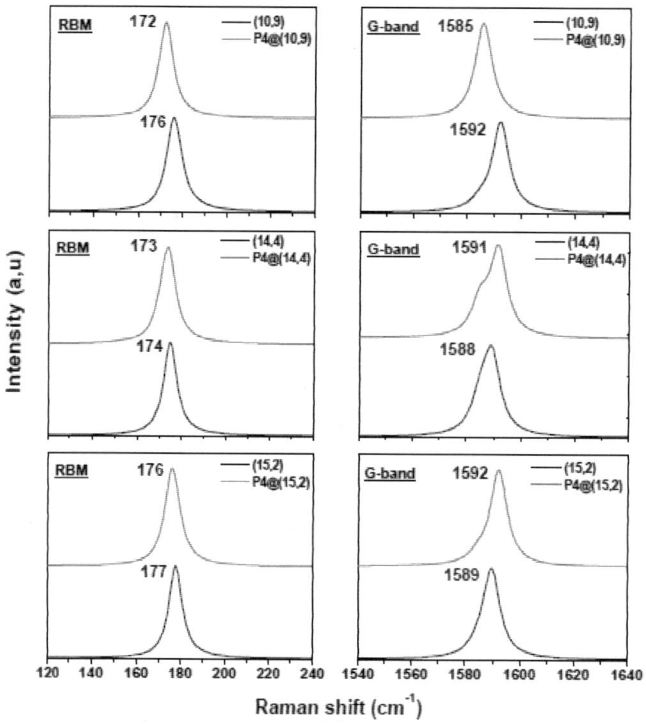

Figure 3. Raman spectra of SWCNTs and PANI4@SWCNTs in the RBM range and G-band.

4. Conclusion

In summary, confinement of PANI4 into SWCNTs was investigated theoretically using Raman spectroscopy. Raman spectrum has been calculated using the spectral moment's method and the optimal configuration of PANI4 in SWCNTs is derived using a convenient Lennard-Jones potential. We found that the optimal diameter calculated for the PANI4@SWCNTs hybrids is around 1.25nm. We have calculated the nonresonant Raman spectra of the obtained configurations corresponding to PANI4@SWCNTs metallic and semiconducting nanotubes. The charges transfer in the

PANI4@SWCNTs systems has been investigated by analysing the behavior of the RBM and G- band modes of SWCNTs before and after encapsulation. The main results suggest a significant charge transfer to and from SWCNTs.

5. References

[1] Lungenschmied C, Dennler G, Neugebauer H, Sariciftci S N, Glatthaar M, Meyer T and Meyer A 2007 *Sol. Energ. Mat. Sol,* **91** 379-384

[2] Green M A, Emery K, Hishikawa Y, Warta W and Dunlop E D 2015 *Prog. Photovoltaics: research and applications,* **23** 1-9

[3] Cheng P and Zhan X 2016 *Chem. Soc. Rev,* **45**, 2544-82

[4] Brabec C, Scherf U and Dyakonov V 2011 *Organic photovoltaics: materials, device physics, and manufacturing technologies.* (John Wiley & Sons)

[5] Ameen S, Shaheer Akhtar M and Husain M 2010 *Sci. Adv. Mater,* **2** 441-462

[6] Hou J, Chen H Y, Zhang S, Chen R I, Yang Y, Wu Y and Li G 2009 J. *Am. Chem. Soc,* **131** 15586-87

[7] Zhao W, Li S, Yao H, Zhang S, Zhang Y, Yang B and Hou J 2017 *J. Am. Chem. Soc,* **139** 7148-51

[8] Sun Y P, Fu K, Lin Y and Huang W 2002 *Acc. Chem. Res,* **35** 1096-1104

[9] Dresselhaus M S, Dresselhaus G, Charlier J C and Hernandez E 2004 *P. Roy. Soc. A-Math Phy,* **362**, 2065-98

[10] Yan J, Ni T, Zou F, Zhang L, Yang D, Yang S and Zou B 2014 *Diam. Relat. Mater,* **41** 79-83

[11] Charekhah R, Jarrahi Z, Darabi M, Imani A and Farzi G 2019 *J. Mater Sc-Mater El,* **30** 26-36

[12] Mandoc M M, Koster L J A and Blom P W 2007 *Appl. Phys. lett,* **90** 133504

[13] Jin S, Jun G H, Jeon S and Hong S H 2016 *Nano. Converg,* **3** 8

[14] Guldi D M and Martín N 2010 *Carbon nanotubes and related structures: synthesis, characterization, functionalization, and applications* (John Wiley & Sons).

[15] Park H, Zhao J and Lu J P 2006 *Nano. Lett,* **6** 916-919

[16] Gao J, Blondeau P, Salice P, Menna E, Bártová B, Hébert C, ... & Loi M A 2011 *Small,* **7** 1807-15

[17] Yanagi K, Miyata Y and Kataura H 2006 *Adv. Mater,* **18** 437- 441

[18] Rahmani A, Sauvajol J L, Rols S and Benoit C 2002 *Phys. Rev. B,* **66** 125404

[19] Soler J M, Artacho E, Gale J D, García A, Junquera J, Ordejón P and Sánchez-Portal D J. 2002 *Phys-Condens. Mat,* **14** 2745

[20] Dresselhaus G, Riichiro S 1998 *Physical properties of carbon nanotubes* (World scientific).

[21] Rahmani A, Sauvajol J L, Rols S and Benoit C 2002 *Phys. Rev. B,* **66** 125404

[22] Rahmani A, Sauvajol J L, Cambedouzou J and Benoit C 2005 *Phys. Rev. B,* **71** 125402

[23] Rahmani A H, Boutahir M, El Biyaali A, Fakrach B, Chadli H, Sbai K,... & Rahmani A 2016 *RSC Adv,* **6** 41025-31

[24] Boutahir M, Rahmani A, Chadli H and Rahmani A 2016 *Eur. Phys. J-Appl. Phys,* **74** 24605

[25] Dresselhaus M. S, Dresselhaus G and Eklund P C 1996 *Science of fullerenes and carbon nanotubes: their properties and applications* (Elsevier)

[26] Malard L M, Pimenta M A A, Dresselhaus G and Dresselhaus M S 2009 *Phys. Rep,* **73** 51-87

[27] Okada S 2007 *Chem. Phys. Lett,* **438** 59-62

[28] Britz D A and Khlobystov A N 2006 *Chem. Soc. Rev,* **35** 637-659

[29] Joung S K, Okazaki T, Kishi N, Okada S, Bandow S and Iijima S 2009 *Phys. Rev. Lett*, **103** 027403

IOP Conf. Series: Materials Science and Engineering **783** (2020) 012023 doi:10.1088/1757-899X/783/1/012023

Enhancement of localized surface plasmon resonances of silver nanoparticles array upon the presence of graphene coatings: LSPR biosensor

Mohamed El Barghouti[1,*], Abdellatif Akjouj[2] and Abdellah Mir[1]

[1]Laboratory of Advanced Materials Studies and Applications (LEM2A), Physics Department, Faculty of Science, Moulay Ismail University, B.P. 11201 Zitoune, Meknes, Morocco

[2]Institute of Electronics, Microelectronics and Nanotechnology, UMR CNRS 8520, Lille University, FST, Department of Physics, 59655 Villeneuve d'Ascq, France

E-mail: *Corresponding author. Mohamed.elbarghouti@gmail.com

Abstract. We study, in this article, the resonance of localized surface plasmon (LSPR) on silver nanoparticles (AgNPs) deposited on a glass substrate and coated with a graphene. The resonance peaks in the absorption spectra of the nanostructure SiO_x/AgNPs/Graphene, show significantly different profiles such as the thickness of the graphene, and the refractive index of the surrounding environment which is, gradually, varied. Numerical simulation reveals that movements in plasmon resonance peaks result from coupling between the AgNPs networks and the covering graphene layer. In addition, the LSPR modes shift to red from 412 nm to 548 nm when the thickness of the graphene layer deposited on silver nanoparticles changes from 0.34 nm to 9 nm. Light coupled with LSPR modes propagating along a AgNPs-graphene interface is calculated and compared to a conventional AgNPs-based LSPR biosensor. The result of the comparison depicts an improvement in the sensitivity of the biosensor from 26 to 106 nm/RIU.

1. Introduction

Optical properties of metal nanoparticles (Ag or Au) and their addressability via spectroscopic techniques have put them in the forefront of fundamental and applied research fields, with applications ranging from nanoparticles array wave guiding to biosensor designs [1-3]. Localized surface plasmonic resonance biosensors (LSPR) are the most advantageous biosensors thanks to their ability to provide biomolecular detection with higher sensitivity, mainly because they eliminate the time-consuming marking process that reduces molecular binding disturbance [4,5]. Understanding their properties is significantly more complex when they are deposited on a substrate, or when they interact [6,7]. It is quite clear from already available theoretical and experimental studies (on metallic nanoparticles) that several parameters do govern their behavior [8,9].

Spectacular progress on the control of optical properties of graphene material sheets stimulated the search for new types of two-dimensional materials. Graphene considered as a very promising new kind of material with possible applications in several domains (including bio-plasmonics). It's well established that AgNPs strongly absorb within narrow frequency bands in the visible range as their localized surface plasmon resonances get excited. Besides, the effect of the dielectric environment has been widely studied in the Refs [10-12]. A part of

IOP Conf. Series: Materials Science and Engineering **783** (2020) 012023 doi:10.1088/1757-899X/783/1/012023

this work consists of investigating the influence of graphene layers (whose refractive index is n_g [13,14]) on AgNPs optical properties. Graphene has aroused considerable interest in recent years because of its distinctive electrical and optical properties [15,16], a highly sensitive LSPR biosensor using the graphene layer to improve the sensitivity caused by the light absorbed by the graphene [10,17].

This is performed in the case when the graphene layers are either deposited on the nanoparticles array. Specifically, we will examine the effect of varying progressively the thickness of graphene layers from 0.34 nm to 9 nm. Different dielectric layers of refractive index n_2 are deposited on top of AgNPs/graphene, which are in turn sitting directly or indirectly on a substrate made of SiO_x (n_1). We study numerically how graphene layers are deposited; those are SiO_x/AgNPs/graphene. The corresponding geometry is well illustrated in Fig. 1. The incident light (along the y-axis) wave whose electric field is polarized along x-axis is also highlighted. Throughout the paper, the geometrical parameters of the structure will be fixed such as the particles length is l = 25nm, a = 70nm is the lattice parameter (along x-axis), and h = 20nm is the nanoparticles height.

Since graphene has important properties, to absorb the biomolecular and, thus, allows a significant change in the refractive index of the graphene-AgNPs interface to protect the structure, which avoids the oxidation of nanoparticles Ag [2]. The numerical results show that, compared to graphene in an LSPR substrate based biosensor/AgNPs, graphene in an LSPR substrate based biosensor/AgNPs/graphene offers a 304.60% higher sensitivity for nine graphene layers.

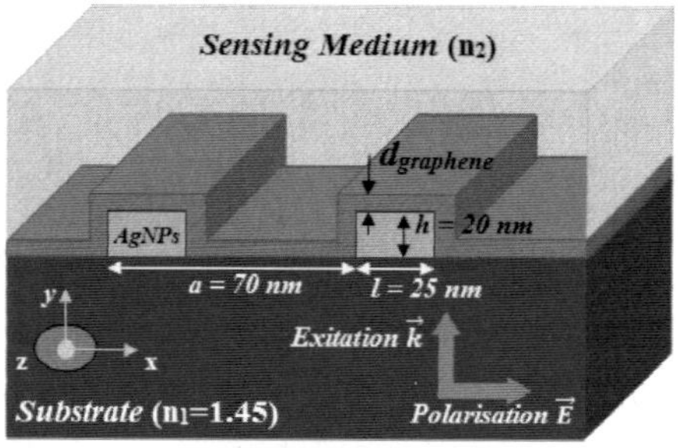

Figure 1. Schematic representation of AgNPs when covered with few layers of graphene and another dielectric lying on top (detection medium of refraction index n_2), and deposited on a SiO_x substrate (SiO_x/AgNPs/graphene/dielectric).

2. Analysis and modelling part

2.1. Simulation model

The optical properties of silver nanoparticles (AgNPs) are determined numerically within a wavelength domain ranging from UV to near IR, specifically, [300 nm-800 nm]. Simulations are performed using Finite Element Method (FEM) [18]. Actually, the simulation method has already been well documented in Refs [19,20]; it focuses on solving Maxwell's equations in the frequency domain by discretizing space in small finite elements (having various possible shapes such as triangular, tetrahedral (3D) etc.). Our calculation is carried out in a two-dimensional (2D) domain with Ox and Oy axes oriented as shown in Fig. 1.

The configuration drawn in Fig. 1 that consists of silver nanoparticles (having length, l, height, h and interparticle distance, a-l) deposited on substrate (of refractive index n_1=1.45), and coated with a graphene layer of the refractive index n_g. This last interface is furthermore covered by a non-absorbing medium of refractive index n_2 such as air and water. For the calculations, the following geometrical parameters were chosen: l=25 nm, h=20 nm, a=70 nm. This corresponds to an average metal coating of 37% [21].

An incident plane wave is generated within SiO_x substrate of refractive index $n_1 = 1.45$, and is launched along the Oy axis (direction of propagation). The plane wave is TM polarized and its associated electric field vibrates along the Ox axis. The calculation is meant to simulate an infinite array of nanoparticles along Ox axis, and the structure is supposed to be infinite along the direction of Oz. The space discretization is realized by means of free triangular mesh and such that the maximum size of elements is adjusted depending on the material considered. In all dielectrics (layer on top of AgNPs and SiO_x substrate), the elements size is no more than $(\frac{\lambda_0}{2n})/5$, whre n is the index of SiO_x or water/air. In Fig. 1, is sketched the simulation models domains in the considered configuration, that is, when nanoparticles are covered with thin layers of graphene. The frequency-dependent complex permittivity of silver (Ag) metal is described by the Lorentz-Drude model [22-24]:

$$\varepsilon(\omega) = \varepsilon_{r,\infty} + \sum_{m=0}^{M} \frac{f_m \omega_p^2}{\omega_m^2 - \omega^2 + j\omega\Gamma_m} \tag{1}$$

where $\varepsilon_{r,\infty}$ is the relative permittivity at infinite frequency, ω_p the plasma frequency, and ω_m, f_m and Γ_m are the resonance frequency, the strength (weight) and the damping rate of the mth oscillator, respectively. The Lorentz-Drude model uses M damped harmonic oscillators to describe the small resonances observed in the metal's frequency response. The value of the dielectric constant at infinite frequency ($\varepsilon_{r,\infty}$=1) and all the other parameters of L-D model are taken from reference Ref. [23].

The graphene deposited between the AgNPs and the detection medium to increase the macromolecular absorption capacity of the biosensor (fixation of the biological element). The thickness of the graphene is varied as $d_{graphene} = L \times 0.34$ nm, where L is the number of graphene layers and its complex refractive index (n_g) in the visible range is an absorbent and dispersing material, the refractive index of which is given by the following Eq. (2) [13]:

$$n_g = 3.0 + i\frac{C_1}{3}\lambda \tag{2}$$

The constant C_1=5.446 μm^{-1} [16], and λ is the wavelength of the incident light in μm. The refractive index of the detection medium is chosen as $n_2 = 1.00$ or 1.333.

2.2. Principle of LSPR sensing

When the dielectric properties of the local environment are modified by the adsorption of analytes to be analysed, there will be a peak wavelength of the LSPR observed shift to red in the absorption spectrum (Fig. 3). Since the refractive index (RI) is related to the dielectric properties, LSPR biosensors can detect the RI changes in the local environment caused by the adsorption of substances to be analysed. The correlation between the peak wavelength displacement ($\Delta\lambda_{LSPR}$) and the RI change (Δn_2) is illustrated as follow:

$$\Delta\lambda_{LSPR} = S\Delta n_2 \tag{3}$$

where S is the refractive index sensitivity of the interface, Δn_2 is the variation of local refractive index of the detection medium [25]. The refractive index sensitivity in nm/RIU (change of nanometer per refractive index unit) is thus defined as,

$$S = \frac{\Delta\lambda_{LSPR}}{\Delta n_2} \tag{4}$$

3. Results and Discussion

In this work, we have studied the detection potential of a 2D structured network of Ag nanoparticles. Through a detailed study using the FEM method, we showed the plasmonic response of the structured presented in Fig. 1. The theoretical performance comparison between structured biosensors and conventional biosensors using a purely propagatifs plasmon, suggests to the importance of nanostructure in enhancing the sensitivity not only in terms to the response of a biosensor without a graphene but also in relation to the impact of nano-structuring of the surface which generates only a loss of performance. We have found that the prevalence of LSPR modes can lead to an increase in terms of sensitivity to a deposit of a graphene layer, while being sensitive to variations of the $d_{graphene}$ with the incidence wave normal to the substrate and resulting in a gain of 304.60%.

In order to determine the effect of graphene layer thicknesses on LSPR of the hybrid nanostructure, the absorption spectrums of the glass/AgNPs/graphene with different graphene layers thicknesses are plotted in Fig. 2. The refractive index (n_2) of the surrounding environment is set to $n_2 = 1.333$ and the graphene layer thickness varies from 0.34 to 9 nm. Each nanostructure presents the plasmonic resonance mode in the absorption spectrum.

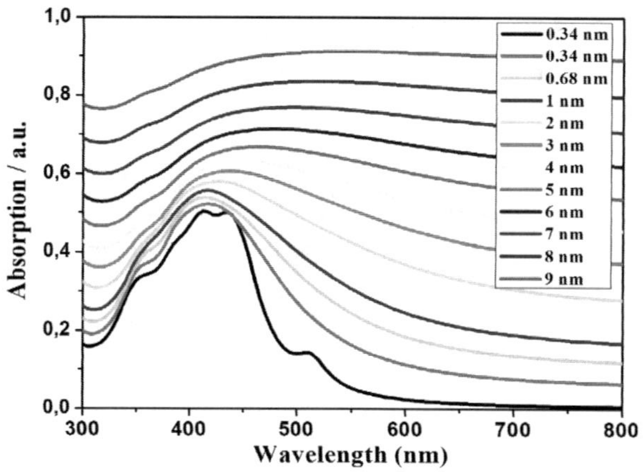

Figure 2. Absorption spectrum of the nanostructure depicted in Fig. 1. The 12 spectrums are given for different values of graphene thickness varying from 0.34 nm to 9 nm.

Numerical results reveals that movements in plasmon resonance peaks result from coupling between the AgNPs networks and the covering graphene layer. In addition, the LSPR modes shift to red from 412 nm to 548 nm. Giving a very significant increase in the resonance wavelength of 136 nm, when the thickness of the graphene layers deposited on Ag nanoparticles varies from 0 to 9 nm.

Specifically, the LSPR in the glass/AgNPs/graphene nanostructure is the convolution of resonance mode; the red-shift is likely due to different dielectric environments probed on the opposite faces of the nanohybrids when the $d_{graphene}$ is varied from 0.34 to 9 nm. It is further believed that the red-shift comes from charge transfer between the graphene layers and Ag nanoparticles array. Therefore, the plasmon resonance on the glass/AgNPs/graphene nanostructures can be tuned by changing the thickness of the graphene layers [10,26].

As the refractive index (RI) of the sample growing, the wavelength also increases. We investigate the sensitivity of the Ag nanoparticles LSPR biosensor in the bio-substances RI range of 1.00-1.333, and the corresponding linear fitting curve of the resonance wavelength with respect to the refractive index is shown in Fig.3a.

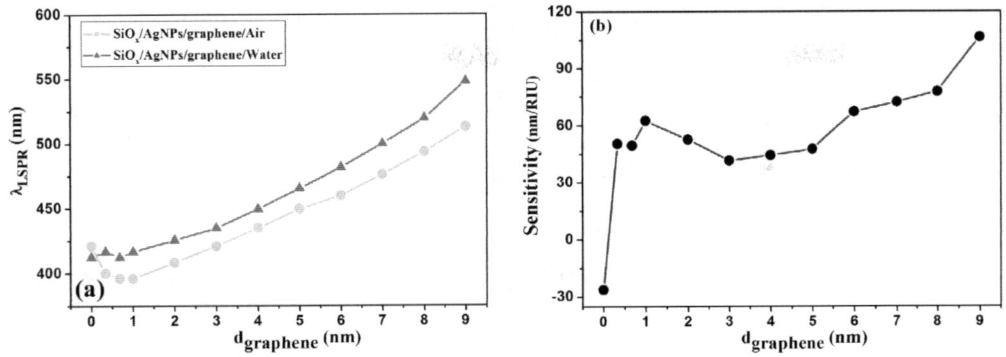

Figure 3. (a) Evolution of the wavelength of LSPR versus the thickness of graphene (b) Variation of the sensitivity versus $d_{graphene}$.

Sensitivity investigation was conducted for graphene layer and biomolecules instead of conventional relatively large number of biomolecules thickness approach without graphene. The sensitivity improved by 304.60% (4.04 times) when the pure AgNPs conventional biosensor was covered with graphene layer of the $d_{graphene}$ = 9 nm.

In this situation, the shift in resonance wavelength (λ_{LSPR}) for the background nanostructure, and increases in a same method for both graphene layer enrichment. This phenomenon can also be seen in Fig. 3a, the shift in resonance wavelength to red, depending on the thickness of dielectric graphene layer ($d_{graphene}$).

Fig. 3b shows that calibration plot of the sensitivity changes with the thickness of graphene. Several numerical simulations were carried out to find how sensitivity is related on the thickness of graphene filed of the nanoparticles. This hybrid nanostructure offers the thinnest design and the better sensing performance than both metal and based approaches, in particular the sensitivity improvement of this design that of pure Ag nanoparticles biosensor. Fig. 3b shows the biosensing performance of the proposed glass/AgNPs/graphene hybrid systems with various graphene layer thicknesses. As indicated, the proposed nanostructure can operate in the visible range, and the sensitivity of detection increases with the thickness of the graphene layer. It is observed, the sensitivity changes when going from $d_{graphene}$ = 0 nm to $d_{graphene}$ = 9 nm.

Adding that in the case of depositing thinner thickness of graphene, there is a strong shift towards the red resonance. Furthermore, it was further explained that the sensitivity of nanosensors based on this phenomenon is related to the shifting of the resonance wavelength of the plasmon response of the nanoparticles with $d_{graphene}$, that is to say in the regions of d where the variation of the $\Delta\lambda_{LSPR}$ is large, the sensitivity is optimal. This indicates a greater effectiveness of bio-detection with these structures.

In a second phase, several numerical simulations were carried out to determine how the sensitivity changes according to the thickness of graphene deposited on the nanoparticles. The results obtained appear in Fig. 3. First, an increase in sensitivity is detected for graphene thicknesses between 0 and 1 nm for nanostructure. In fact, (Fig.3b) S = 26.30 nm/RIU for $d_{graphene}$=0 nm and S = 62.50 nm/RIU for $d_{graphene}$=1 nm, corresponding to an increase of about 137.60%. Then the sensitivity decreases from $d_{graphene}$=1 nm, to reach 41.61 nm/RIU when $d_{graphene}$=3 nm.

On the other hand, for ($d_{graphene}$), either when the nanoparticles are covered by different graphene slips $3nm \leq d_{graphene} \leq 9nm$ or when the AgNPs are covered with 27 layers of graphene ($d_{graphene}$ = 9 nm), the resonance wavelength is λ_{LSPR} = 512.82 nm for n_2=1.00 and λ_{LSPR} = 547.94 nm for n_2=1.333, which corresponds to a sensitivity of S = 106.44 nm/RIU.

First of all, there has been an increase in sensitivity (glass/AgNPs/graphene) for thicknesses of graphene between 3 and 9 nm. Indeed, S = 41.61 nm/RIU for 3 nm and S = 106.44 nm/RIU to 9 nm.

Fig. 4 shows the electric field strength of the nano-system at different thicknesses from the graphene layers to the surface plasmon resonance.

The excited electrical field distributions on plasmon resonance, at a normal incidence structures, are represented by cards of electric field in Fig. 4, which indicates that this mode is a dipolar localized surface plasmon, whose hot spots (areas of high field strengths) are pushed to the top and lower corners of the AgNPs of the structure lacking the graphene layer (see Fig. 4a), and to lower corners of higher AgNPs graphene layer.

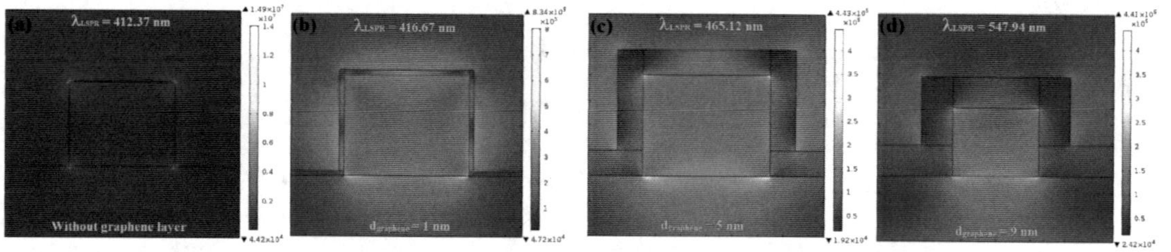

Figure 4. Maps of electric field in glass/AgNPs/graphene structure for $d_{graphene}$ = 0nm under monochromatic incident at 412.37nm (a), for $d_{graphene}$ = 1nm at 416.67nm (b), for $d_{graphene}$ = 5nm at 465.12nm (c) and for $d_{graphene}$ = 9nm at 547.94nm (d).

AgNPs/graphene based sensor exhibits efficient charge transfer due to low optical absorption efficiency (about 2.3%) of graphene [27]. Generous electron transfer from graphene to Ag nanoparticle will occur in case of optical excitation as the work function of these materials are related as graphene (4.5eV) $<$ Ag (4.74eV) [28,29]. This process will lead to a larger electric field enhancement at the sensing inter-face thereby resulting in a higher sensitivity to the target biomolecules [30].

For the nanostructure SiO_x/AgNPs/graphene in [Fig. 4b, c and d] the fields are, basically, limited to the space between the graphene layer and the dielectric of detection which are lower than the AgNPs. It can be seen by comparing Fig. 4a that when the nanoparticles without graphene layer have hot spots at each of their corners.

Moreover, adding 1 nm of graphene has resulted in an increase in the amplitude of the electric field in the interparticle space followed by a decrease in the peak value intensity. This observed decrease in field strength is due to the coupling between the plasmon mode and the Fabry-Perot cavity modes connected to the presence 5 and 9 nm of the graphene layer and sensing medium presented above the Ag nanoparticles [2,30].

The absorption spectra of the glass/AgNPs/graphene nanostructure for different refractive indexes of the surrounding medium are shown in Fig. 5a. A series of refractive indexes n_2 = 1; 1.333; 1.45; 1.6; 2; 2.5 and 3 are used in the calculation.

The red-shift in the plasmonic mode from 449.44 to 597 nm bears a positive relationship with the refractive index between 1 and 3. This variation is due to the change in the refractive index of the surrounding medium around the glass/AgNPs/graphene nanostructure, resulting in changes in the resonance condition and location of the resonance modes [5,10]. The geometric parameters of the Ag nanoparticles are: l = 25 nm, h = 20 nm, a = 70 nm and $d_{graphene}$ = 5 nm.

This phenomenon can also be seen in Fig. 5b, the change in resonance wavelength shifts to the refractive index of the sensing medium (n_2). Fig. 5b shows that calibration curve is linear, and graphene/AgNPs hybrid nanostructure provides the thinnest design and the better sensing

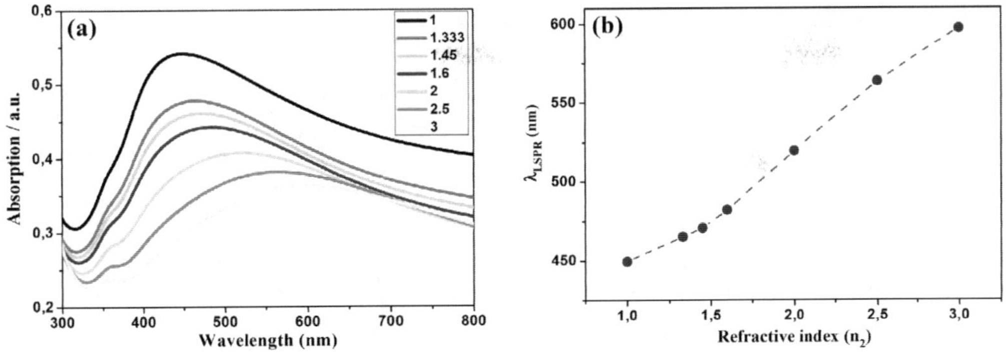

Figure 5. (a) Absorption spectrum corresponding to the structure glass/AgNPs/graphene, for defeated refractive index of the detection medium. (b) Evolution of the resonant wavelength of the LSPR versus refractive index of the detection medium (n_2).

performance than both metal and graphene based approaches, in particular the sensitivity enhancement of this design is 4.04 times larger than that of pure metal nanoparticles biosensor.

4. Conclusions

This paper presents the designing of graphene coated two types of biosensor. The wavelength interrogation mode for setting up localised surface plasmon resonance condition is used to analyze the proposed LSPR biosensors. The spectral study to detect for the biological elements using a compatible sensitive medium. Silver nanostructures (AgNPs) on the glass and covered with a thin dielectric layer of graphene, indicates that this study allows a better insight of systems based on resonant metal nanoparticles coupled with graphene layers. To confirm the model, we compared the resulting spectrum obtained through numerical simulations of LSPR to those released without a layer of graphene. From a practical standpoint, it opens the door to nanotechnology in a controlled and predictable manner. The spectral properties of systems based on Ag-graphene nanoparticles to enhance their particular applicability for sensing applications. This study opens the way to highly sensitive sensors and improved biosensors.

References

[1] Fano U. (1941) The Theory of Anomalous Diffraction Gratings and of Quasi-Stationary Waves on Metallic Surfaces (Sommerfelds Waves). *J. Opt. Soc. Am.* 31, 213.

[2] Reed J C, Zhu H, Zhu A Y, Li C, and Cubukcu E. (2012) Graphene-enabled silver nanoantenna sensors, *Nano letters* 12, 4090-4094.

[3] El Barghouti M, Akjouj A and Mir A. (2018) Effect of MoS_2 layer on the LSPR in periodic nanostructures, *Optik* 171, 237-246.

[4] Homola J and Piliarik M. (2006) Surface Plasmon Resonance Based Sensors, *Springer* 4, 46-47.

[5] El Barghouti M, Mir A, Akjouj A and Szunerits S. (2019) Modulation of localized surface plasmon resonances of a silver nanoparticle array upon the presence of MoS_2 coatings or underlying thin films, *Optik* 179, 819-827.

[6] Nikitin A G, Nguyen T, and Dallaporta H. (2013) Narrow plasmon resonances in diffractive arrays of gold nanoparticles in asymmetric environment: Experimental studies, *Appl. Phys. Lett.* 102, 221-116.

[7] Kretschmann E and Raether H. (1968) Notizen: Radiative decay of nonradiative surface plasmons excited by light, *Z. Naturforsch. A.* 23, 2135-2136.

[8] Galopin E, Noual A, Jonsson J N, Martin J N, Akjouj A, Pennec Y, Rouhani B D, Boukherroub R and Szunerits S. (2009) Short- and Long-Range Sensing Using Plasmonic Nanostructures: Experimental and Theoretical Studies, *J. Phys. Chem. C* 113, 15921-15927.

[9] Touahir L, Galopin E, Boukherroub R, Gouget-Laemmel A C, Chazalviel J N, Ozanam F, Saison O, Akjouj A, Pennec Y, Djafari-Rouhani B, and Szunerits S. (2011) Plasmonic properties of silver nanostructures coated

with an amorphoussilicon-carbon alloy and their applications for sensitive sensing of DNA hybridization, *J. Phys.: Analyst* 136, 1859.
[10] El Barghouti M, Akjouj A and Mir A. (2018) Effect of graphene layer on the localized surface plasmon resonance (LSPR) and the sensitivity in periodic nanostructure, *Photon. Nanostruct. Fundam. Appl.* 31, 107-114.
[11] Malinsky M D, Kelly K L, Schatz G C and Van Duyne R P. (2001) Chain Length Dependence and Sensing Capabilities of the Localized Surface Plasmon Resonance of Silver Nanoparticles Chemically Modified with Alkanethiol Self-Assembled Monolayers, *J. Am. Chem. Soc.* 123, 1471-1482.
[12] El Barghouti M, Akjouj A and Mir A. (2018) Numerical analysis of MoS_2 on the LSPR in periodical nanostructure of Ag arrays/graphene for sensitive sensors, *Int. J. Microwave Opt. Technol.* 13, 454-461.
[13] Bruna M and Borini S. (2009) Optical constants of graphene layers in the visible range, *Appl. Phys. Lett.* 94, 213116.
[14] Fang Z, Wang Y, Liu Z, Schlather A, Ajayan P M, Koppens F H L, Nordlander P and Halas N J. (2012) Plasmon-Induced Doping of Graphene, *ACS Nano.* 6, 10222-10228.
[15] Wu J, Zhou C, Yu J, Cao H, Li S and Jia W. (2014) Design of infrared surface plasmon resonance sensors based on graphene ribbon arrays, *Optics and Laser Technology* 59, 99-103.
[16] Castro Neto A H, Guinea F, Peres N M R, Novoselov K S, and Geim A K. (2009) The electronic properties of graphene, *Rev. Mod. Phys.* 81, 109-162.
[17] Wu L, Chu H S, Koh W S, and Li E P. (2010) Highly sensitive graphene biosensors based on surface plasmon resonance, *Opt. Exp.* 18, 14395-14400.
[18] Dossou K, Byrne M A and Botten L C. (2006) Finite element computation of grating scattering matrices and application to photonic crystal band calculations, *Journal of Computational Physics* 219, 120-143.
[19] Knight M W and Halas N J. (2008) Nanoshells to nanoeggs to nanocups: optical properties of reduced symmetry core-shell nanoparticles beyond the quasistatic limit, *New J. Phys.* 10, 105006.
[20] Hugonin J P, Besbes M and Lalanne P. (2008) Hybridization of electromagnetic numerical methods through the G-matrix algorithm, *Optics Letters* 33, 1590-1592.
[21] Qiang R, Chen R L and Chen J. (2004) Modeling electrical properties of gold films at infrared frequency using FDTD method, *J. Infrared Millimeter Waves* 25, 1263.
[22] Johnson P B and Christy R W. (1972) Optical constants of the noble metals, *Physical Review B* 6, 4370-4379.
[23] Palik E D. (1985) Handbook of Optical Constants of Solids.
[24] Zeng S and al. (2011) A review on functionalized gold nanoparticles for biosensing applications, *Plasmonics.* 6, 491-506.
[25] Anker J N, Hall W P, Lyandres O, Shah N C, Zhao J and Van Duyne R P. (2008) Biosensing with plasmonic nanosensors, *Nature Materials* 7, 442-453.
[26] Saison-Francioso O, Lévêque G, Akjouj A, Pennec Y, Djafari-Rouhani B, Szunerits S and Boukherroub R. (2012) Plasmonic nanoparticles array for high-sensitivity sensing: a theoretical investigation, *J. Phys. Chem. C* 116, 17819–17827.
[27] iovannetti G, Khomyakov P A, Brocks G, Karpan V M, Van den Brink J and Kelly P J. (2008) Doping graphene with metal contacts, *Phys. Rev. Lett.* 101, 026803.
[28] Sachs B, Britnell L, Wehling T O, Eckmann A, Jalil R, Belle B D, Lichten-stein A I, Katsnelson M I and Novoselov K S. (2013) Doping mechanisms in graphene-MoS_2 hybrids, *Appl. Phys. Lett.* 103, 251607.
[29] Zeng S, Hu s, Xia J, Anderson T, Dinh X Q, Meng X M, Coquet P and Yong K T. (2015) Graphene-MoS_2 hybrid nanostructures enhanced surface plasmon resonance biosensors, *Sens. Actuators B: Chem.* 207, 801–810.
[30] El Barghouti M, Akjouj A and Mir A, (2019) Reticular plasmon resonance detection properties of metal nanoparticles, *Physica E: Low-dimensional Systems and Nanostructures* 110, 107-114.

IOP Conf. Series: Materials Science and Engineering **783** (2020) 012024 doi:10.1088/1757-899X/783/1/012024

COMPUTATIONAL EFFECTIVE THERMAL CONDUCTIVITY OF POLYURETHANE MIXED CELL FOAMS

Chaïmaâ HERMAMA[1,*], Anas ELMALIKI[1], and Salma LAHBABI[2]
[1]Equipe de Recherche Appliquée sur les Polymères (ERAP), ENSEM, Université Hassan II, Morocco
[2]Equipe sécurité des systèmes et modélisation, (ESSM), Laboratoire de recherche en ingénierie, ENSEM, Université Hassan II, Morocco
[*] Corresponding author: chaimaa.hermama@ensem.ac.ma

Abstract. The objective of this paper was to calculate effective thermal conductivity of polyurethane foam used in thermal insulation of buildings with a new modeling approach. The proposed approach was more realistic as it simulated the real fraction of closed cells and open cells in the foam and it modeled the real physical phenomena that happen when both types of cells are present. This study was investigated for polyurethane foam with 70 % of closed cells and 30 % of open cells, by using finite element method and numerical homogenization. The result showed that there was a systematic change in thermal conductivity when the type of cell (closed, open and mixed cell) varied at fixed volume fraction. Also the effective thermal conductivity of mixed cell of this PU foam was about λ_m= 0.07 W.m-1.K-1. Simulation proved the interest of this approach. Indeed, it brought new factor that influence the effective parameters which was fraction of closed cells and open cells. They suggested new method for computing thermal conductivity as a function of thermal conductivity of closed and open cell foam.

Nomenclature

QUANTITY	SYMBOL	SI UNIT
Density of PU foam, solid matrix and air	ρ, ρ_s, ρ_g	Kg/m^3
Porosity	ξ	%
Solid volume fraction	f_v	-
Fraction of closed cell	f_r	-
Length of the struts	l	m
Diameter of the strut	d	m
Directional nodal Temperature	Ti, i=x,y,z	K
Directional nodal heat flux	qi, i=x,y,z	$W/m2$
heat conductivity	$\lambda, \lambda_m, \lambda_o, \lambda_c$	$W.m^{-1}.K^{-1}$
Total volume, volume of solid matrix, volume of air	V, V_s, V_g	m^3

1. Introduction

The material behavior is determined by its characteristics parameters, balance equation and its constitutive laws. The coefficients such as the thermal conductivity, permeability and Young's module are intrinsic to the material and help to identify the material application. This macroscopic description can be obtained either by experiment, or by theory. In the case of heterogeneous material, the experimental characterization cannot give intrinsic coefficients because this kind of study is limited to the continuous medium [1]–[3], and experimental conditions impact the results. Analytical methods are appropriate to the case of heterogeneous materials having a simple microstructure like composite [4]. This analytical description is delicate in the case of porous materials, because they have a random geometry structure (distribution, orientation, shape and volume fraction). Unlike theoretical and experimental studies, numerical ones overcome those problems, so we can reconstruct the complex morphology and compute effective coefficients of this kind of materials using numerical simulation.

The geometry structure of the pores is the main parameter affecting the properties of heterogeneous material. KANAUN [5] [6] studies the effect of struts and cell size distribution on effective thermal conductivity of metallic open cell foam using computational homogenization. The numerical analysis of effective thermal conductivity was investigated for plastic closed cell foams in [7]. At fixed values of porosity and number of struts per cell of open cell foam, if the size of cells increase, radiative conductivity increases as well [8]. The influence of microstructure also appears on the properties of convection for open cell foam, as the number of Nusselt increases simultaneously with increasing porosity at constant cell density [9] [10]. Also Marvi-Mashhadi an al. [11] [12] study mechanical behavior of polyurethane closed cell foams by means of computational homogenization.

Most of geometrical modeling of metallic or plastic foams in the literature consider only open cells or closed cells. However, current manufacturing processes cannot reach foams with 100% of only closed cells [13], also in order to have combined characteristics of both types of foam (for example thermal and acoustic insulation) it is necessary to introduce both open and closed cells. Also According to the fraction of closed cell, PU foam is classified in 4 categories [14] CCC1($fr < 20\%$), CCC2($20\% \leq fr \leq 80\%$), CCC3 ($80\% < fr \leq 89\%$) and CCC4($fr \geq 90\%$). Therefore we propose to model the structure of the foam with the real fraction of closed cells and open cells.

2. ANALYSIS AND MODELLING

2.1. Geometrical modeling of the microstructure

Polyurethane foam is a porous heterogeneous material; the microscopy parameters vary in the material. It is made of thin struts which compose the solid phase and cavities. The cavities are polyhedral cells with polygonal faces (in the case of closed cells) and edges.

In this study, the geometry of the polyurethane foam is considered as a periodic structure and characterized by a representative elementary volume (REV). The construction of the REV is based on the kelvin cell method [15] and has a tetrakaidecahedron shape. This method is an idealization of geometrical modeling of the microstructure [16]. Three types of microstructures are considered:

- Microstructure 1: with open cell only
- Microstructure 2: with closed cell only
- Microstructure 3: with mixed cells

Figure 1 illustrates Kelvin multi-cells models in the three cases for 20 cells.

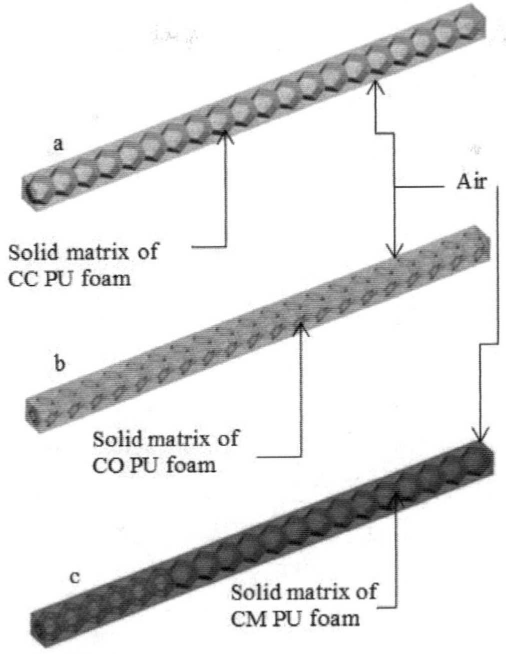

Figure 1 Geometrical modeling of PU foam using 20 cells: a. Closed cells PU foam (CC), b. Open cells PU foam (CO), c. Mixed cells PU foam (CM) with 70 % of closed cells and 30 % of open cells.

The parameters of geometrical modeling of the kelvin cells (the length (l) and the width (d) of the struts (see Figure 2) are taken from the morphology analysis of the microstructure of PU foam in [17]. The fraction considered of closed cell is 70%, the diameter of ligament is about 48 μm and the length is about 160 μm. The experimental porosity of this material is 98% [17].

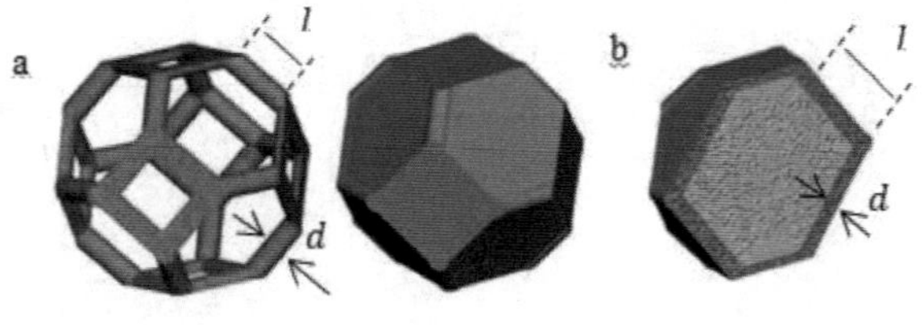

Figure 2 Geometry parameters of single Kelvin cells: a. open cell, b. closed cell

2.2. Homogenization theory

The homogenization theory was introduced by Enrique Sanchez-Palencia. It assumes the existence of homogeneous material which is macroscopically equivalent to the heterogeneous material. It is continuous medium that behaves in "Average" as the heterogeneous material [18]. So thermal conductivities of PU foams can be determined by solving Fourier's law:

$$\langle \vec{q} \rangle = -\langle \lambda . \vec{\nabla} T \rangle \quad (1)$$

The Hill-Mandel criteria [19] give the equivalence of the average heat dissipation and the heat dissipation of the averages:

$$-\langle \vec{q}.\vec{\nabla}T \rangle = -\langle \vec{q} \rangle \langle \vec{\nabla}T \rangle = \langle \vec{q} \rangle \langle -\vec{\nabla}T \rangle \ (2)$$

The assumption of the unique equivalent continuous medium introduces thermo-physical parameters associated to the medium porous and defined as macroscopic effective coefficients. So the effective thermal conductivity $\langle \lambda \rangle$ corresponding to the average of the local value at P $\lambda(P)$ inside the REV:

$$\langle \lambda \rangle = \frac{1}{REV} \int_{REV} \lambda(P) dV \ (3)$$

2.3. Computational thermal simulation

The computation of the effective thermal conductivity of PU foam requires studying thermal conduction in the direction of the thickness. So we apply Uniform Dirichlet conditions in both boundaries: T_A= 277,15K & T_B=317,15K in order to generate conduction. Pure polyurethane and air have a constant thermal conducting in this temperature range: the considered thermal conductivity for matrix phase is 0.29 $W.m^{-1}.K^{-1}$ and for pores 0.024$W.m^{-1}.K^{-1}$. The value used are taken from [20].

The technique used for meshing is finite element method with unstructured meshing. It is the most popular method for discretization of complex geometry in the case of linear problem as heat equation in the case of steady- state conduction.

For that purpose microstructure parameters are unchanged, but different number of cells (1, 10, 20 and 30 cells) and different mesh resolutions are used in the three cases open, closed and mixed cell types.

The parameters of PU foam are deduced from calculating directional heat flux q_x, q_y, q_z (Figures 3,4,5 Directional heat flux for 20 mixed cell) and directional temperature gradient $\nabla_x T$, $\nabla_y T$, $\nabla_z T$ (Figures 6,7,8 directional temperature gradient for 20 mixed cell) over the nodes of the domain by using ANSYS workbench, then computing the averages.

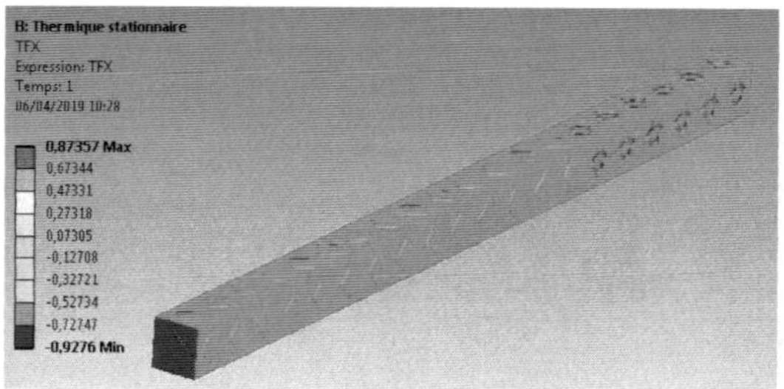

Figure 3 Directional heat flux q_x

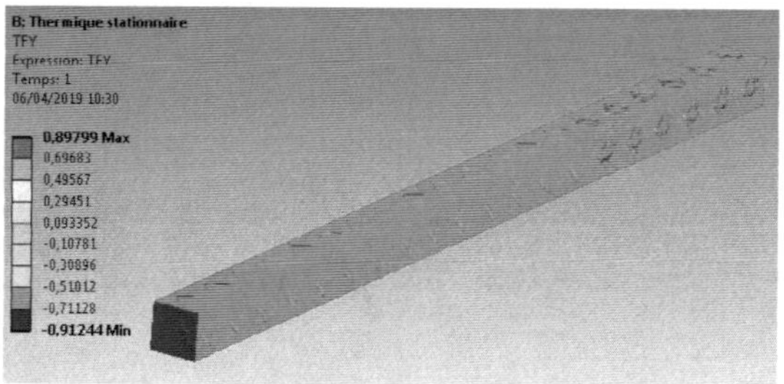

Figure 4 Directional heat flux q_y

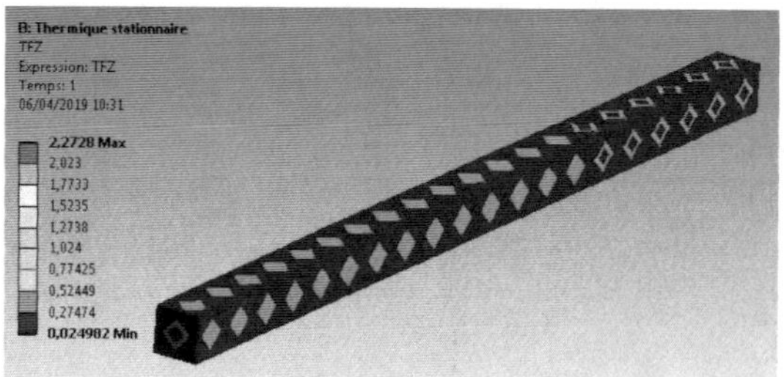

Figure 5 Directional heat flux q_z

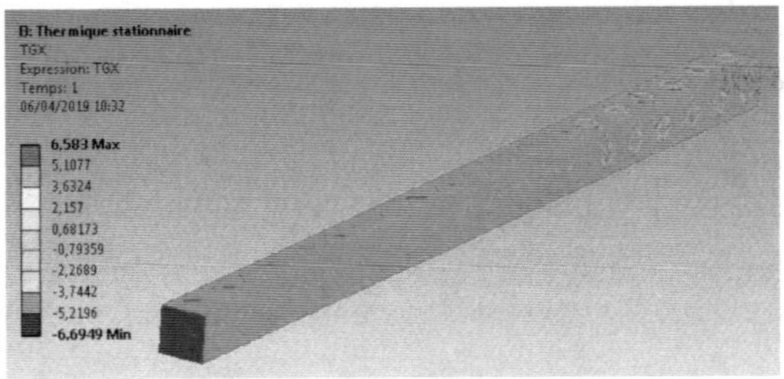

Figure 6 Directional temperature gradient $\nabla_x T$

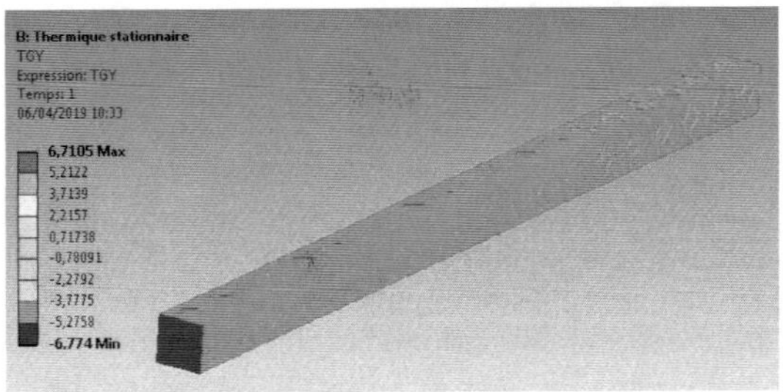

Figure 7 Directional temperature gradient $\nabla_y T$

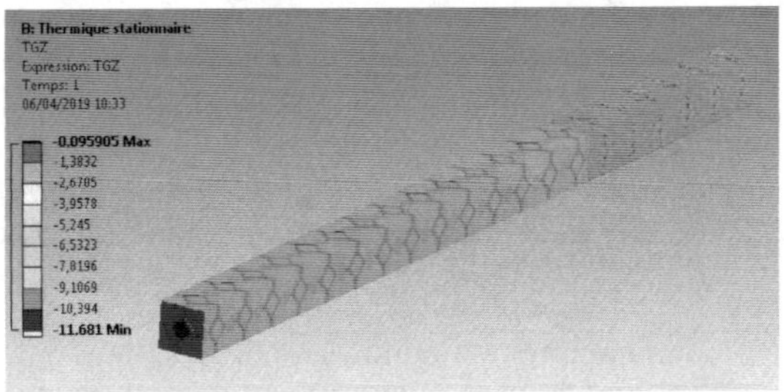

Figure 8 Directional temperature gradient $\nabla_z T$

So if we assume that the macroscopic equivalent is an isotropic and homogeneous material, the effective thermal conductivity is giving by [21]:

$$\lambda = -\frac{\langle q_x \rangle \langle \nabla_x T \rangle + \langle q_y \rangle \langle \nabla_y T \rangle + \langle q_z \rangle \langle \nabla_z T \rangle}{\langle \nabla_x T \rangle^2 + \langle \nabla_y T \rangle^2 + \langle \nabla_z T \rangle^2} \quad (4)$$

3. Validation of the geometrical modeling and validation OF the RESULTS

In order to confirm precision relevance suitability of geometrical modeling the PU foam, we compare apparent porosities of the model with the theoretical and the experimental ones. There are two methods of computing porosity ξ: the first one is established the relation between porosity, the density of the PU foam ρ and the density of the pure polyurethane ρ_s:

$$\xi = 1 - \frac{\rho - \rho_s}{\rho_s - \rho_g} \approx 1 - \frac{\rho}{\rho_s} \; si \; \rho \gg \rho_g (5)$$

Gibson and Ashby approaches [16], gives a density of a package of Kelvin cells as a function of the struts dimensions, to compute theoretical porosities in the 3 cases :

- Theoretical porosity of a tetrakaidecahedron open cell :

$$\xi_o = 1 - 1.06 * \frac{d^2}{l^2} (6)$$
$$\xi_o = 90.46\%$$

- Theoretical porosity of a tetrakaidecahedron closed cell :

$$\xi_c = 1 - 1.18 * \frac{d}{l} \; (7)$$
$$\xi_c = 64.6\%$$

- And we can find that theoretical porosity of a tetrakaidecahedron mixed cell :

$$\xi_m = (1 - f_r) * \xi_o + f_r * \xi_c \; (8)$$
$$\xi_m = 72.36\%$$

The second method gives the porosity ξ as a function of the solid fraction. It is used to compute apparent porosities of the models:

$$\xi = 1 - f_v = \frac{V_g}{V} \quad (9)$$

So we evaluate apparent porosity for the different models and we calculate the relative error. We notes that it is lower than 16% in comparing with theoretical values, and it is lower than 18% in comparing with the experimental ones (see Table 1 for more details).

Table 1
Comparison between Apparent, Experiment and theoretical porosity

Cell types	Apparent porosity	Relative error compared to the experimental	Relative error compared to the theory
Closed cell	74.9%	-	16 %
Open cell	93.6 %	-	3.4 %
Mixed cell (fr=70%)	80 %	18%	11%

Thermal erreor is an important parameter to check the validation of results, it evaluate the continuity of flux and temperature gradient. High value requires refinement of the mesh. It is based on error estimation technique for adaptive finite element analysis of heat conduction problems given by Huang and Lewis [22].
Thermal erreor for each element i is [23] :

$$e_i = \frac{1}{2}\int_{V_e} \{\Delta q\}^T [D]^{-1} \{\Delta q\} dV_e \quad (10)$$

Where :
e_i: energy error for element
V_e: volume of element
$[D]$: conductivity matrix
$\{\Delta q\}$: thermal flux error vector

The thermal error of geomerical modeling of PU foam at converged value of thermal conductivity are reported in Table 2. Figure 9 show exemple of thermal error of 20 mixed cell PU foam.

Table 2
Thermal error at converged value of thermal conductivity

Cell types	Number of cells	Average thermal error
Closed cell	1	7.04E-10
	10	2.1975
	20	53.6901
	30	9.79E-09
Open cell	1	1.75E-08
	10	38.1283
	20	8.83E-09
	30	4.46E-09
Mixed cell (fr=70%)	10	25.4981
	20	1.18E-08
	30	23.0786

IOP Conf. Series: Materials Science and Engineering **783** (2020) 012024 doi:10.1088/1757-899X/783/1/012024

Figure 9 thermal error of 20 mixed cell PU foam

4. RESULTS AND DISCUSSION

In this part, we verify the convergence of effective thermal conductivities. The results should be more accurate when mesh density and/or number of cells increases, so the results converge to a single value.

The numerical results for the three cases are compared with different analytical bounds: Hashin–Shtrikman bounds (H.S bound+ & bound-), and models such as: series model, parallel model, Self-consistent model (S.C) and the two forms of Maxwell-Eucken models (M.E model 1 & 2) (Figures 10, 11 and 12). These bounds and models don't consider distribution, shape, and orientation of cells. The only morphological parameter used is volume fraction. For computing the analytical effective thermal conductivity, we use the forms described in [4], [21], And the apparent volume fraction.

The analytical models and bounds used for thermal analysis are given by following equations:

- Series model: $\lambda = \frac{\lambda_s \lambda_g}{(1-f_v)\lambda_s + f_v\,\lambda_g}$ (11)
- Parallel model: $\lambda = f_v\,\lambda_s + (1-f_v)\lambda_g$ (12)
- Self-consistent model:

$$\lambda = \frac{1}{4}\left[(3f_v - 1)\lambda_s + \lambda_g(3(1-f_v) - 1) + \sqrt{[(3f_v - 1)\lambda_s + \lambda_g(3(1-f_v) - 1)]^2 + 8\lambda_s\lambda_g}\right] \quad (13)$$

- Maxwell-Eucken model 1: Air considered as continuous phase:
  $\lambda = \lambda_g \frac{2\lambda_g + \lambda_s - 2(\lambda_g - \lambda_s)f_v}{2\lambda_g + \lambda_s + (\lambda_g - \lambda_s)f_v}$ (14)
- Maxwell-Eucken model 2: pure polyurethane considered as continuous phase: :
  $\lambda = \lambda_s \frac{2\lambda_s + \lambda_g - 2(\lambda_s - \lambda_g)(1-f_v)}{2\lambda_s + \lambda_g + (\lambda_s - \lambda_g)(1-f_v)}$ (15)
- Hashin–Shtrikman bound +:
  $\lambda = \lambda_g + \frac{f_v}{\frac{1}{\lambda_s - \lambda_g} + \frac{1-f_v}{3\lambda_g}}$ (16)
- Hashin–Shtrikman bound -:
  $\lambda = \lambda_s + \frac{1-f_v}{\frac{1}{\lambda_g - \lambda_s} + \frac{f_v}{3\lambda_s}}$ (17)

The convergence value of effective thermal conductivity of closed cell of this PU foam is about λ_c =0.08 $W.m^{-1}.K^{-1}$ (Figure 10). The numerical results are approximately between parallel model and H.S bound- (which is equal to M.E model 2).

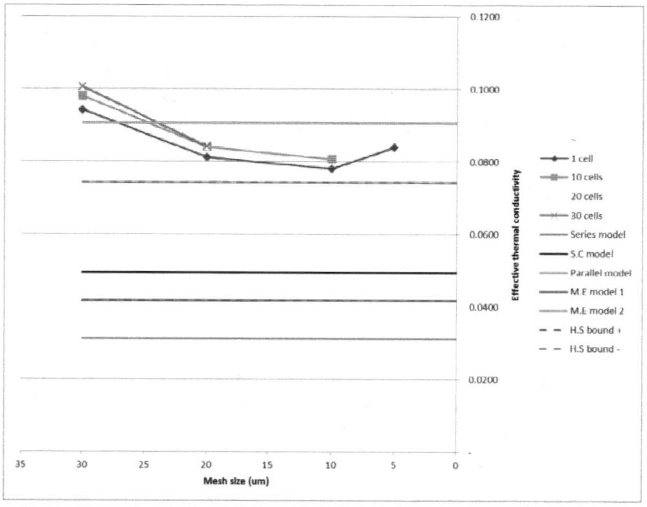

Figure 10 Effective thermal conductivity of closed cell PU foam ($W.m^{-1}.K^{-1}$) versus Mesh size (um).

The convergence value of effective thermal conductivity of open cell of this PU foam is about λ_o =0.05 $W.m^{-1}.K^{-1}$ (Figure 11). The numerical results are close to parallel model.

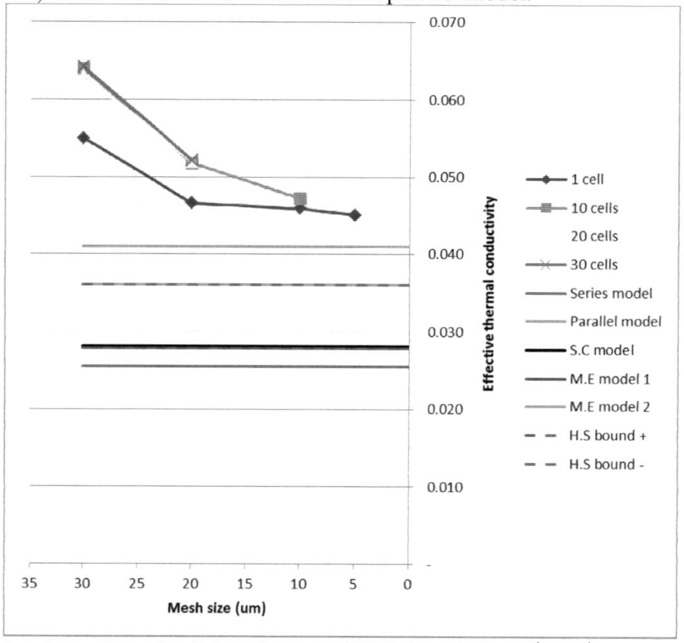

Figure 11 Effective thermal conductivity of open cell PU foam ($W.m^{-1}.K^{-1}$) versus Mesh size (um).

The convergence value of effective thermal conductivity of mixed cell of this PU foam is about λ_m =0.07 $W.m^{-1}.K^{-1}$ (Figure 12). The numerical results are approximately between parallel model and H.S bound- (which is equal to M.E model 2).

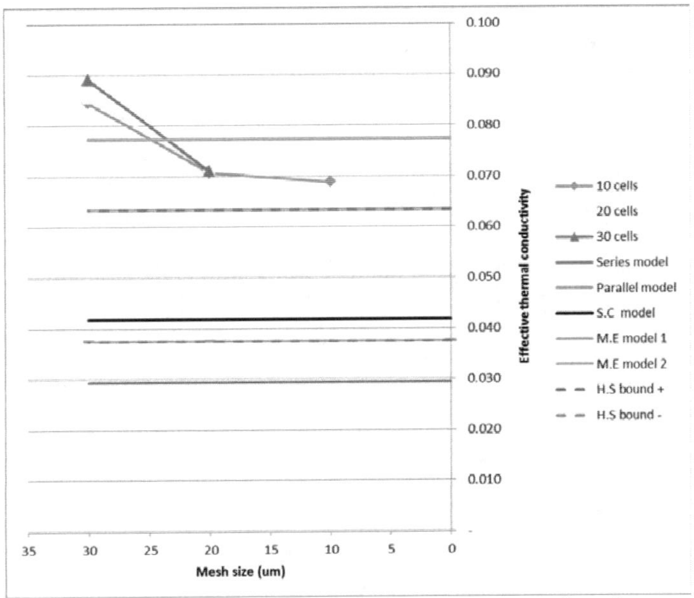

Figure 12 Effective thermal conductivity of mixed cell PU foam ($W.m^{-1}.K^{-1}$) versus Mesh size (um).

- Thermal conductivity converge rapidly for finer meshes and large volumes for the three cases.
- We can perceive that the thermal conductivity of mixed cells (which represents the real morphology of the PU foam) is a linear function of thermal conductivity of open and closed cells:

$$\lambda_m = \lambda_o(1 - f_r) + \lambda_c . f_r$$

- The numerical results are based on the assumption of averaging values of thermal conductivities, these averages are computing in nodes which describe the geometry of the material and this explains the reason that the closest analytical model to the numerical results in the three cases (close, open and mixed cell) is parallel model.
- Maxwell-Eucken model 2 is also close to the results because it includes the distinction between the dispersed phase and the continuous phase and this is the reason make Maxwell-Eucken model 1 is far from the numerical results.
- The analytical homogenization of random heterogeneous materials is introduced by Kroener [24]. Self-consistent model is used when the material has perfectly random material and anisotropic. Hashin–Shtrikman bounds are used in the case of random and isotropic material [25]. The materiel studied is considered as periodic and isotope. So the difference between numerical results and analytical ones maybe caused by this reasons.

CONCLUSIONS

This study is a computational method for calculating thermal conductivity of PU foam with the new approach. Two ways investigated for computing effective thermal conductivity of PU mixed cell foam: First method, the effective thermal conductivity of mixed cell foam is a linear function of thermal conductivity of open and closed cell foam. Second method is to compute directly the effective thermal conductivity from the model of mixed cell foam.

This study does not consider some material characteristic such as the aspect of randomly and anisotropic, so in order to develop the numerical model of the new approach we must include these parameters.

ACKNOWLEDGMENTS

L'Ecole Normale Supérieure de L'Enseignement Technique de Rabat is acknowledged with thanks. The authors' special thanks go to Dr. Farah abdoun a professor at L'Ecole Normale Supérieure de L'Enseignement Technique de Rabat, who give us the access to ANSYS workbench.

References

[1] "ISO 22007-2 Plastiques - Détermination de la conductivité thermique et de la diffusivité thermique - Partie 2: Méthode de la source plane transitoire (disque chaud)," *Plast. -- Déterm. Conduct. Therm. Diffus. Therm. -- Partie 2 Méthode Source Plane Transit. Disq. Chaud.*

[2] "ISO 22007-3 Plastics -Determination of thermal conductivity and thermal diffusivity -Part 3: Temperature wave analysis method," *Plast. -- Determ. Therm. Conduct. Therm. Diffus. -- Part 3 Temp. Wave Anal. Method.*

[3] "ISO 22007-4 Plastics - Determination of thermal conductivity and thermal diffusivity - Part 4: Laser flash method," *Plast. -- Determ. Therm. Conduct. Therm. Diffus. -- Part 4 Laser Flash Method.*

[4] Progelhof R C, Throne J L, and Ruetsch R R, "Methods for predicting the thermal conductivity of composite systems: A review," *Polym. Eng. Sci.*, vol. 16, no. 9, pp. 615–625, Sep. 1976, doi: 10.1002/pen.760160905.

[5] Kanaun S and Kochekseraii S B, "Conductive properties of foam materials with open or closed cells," *Int. J. Eng. Sci.*, vol. 50, no. 1, pp. 124–131, Jan. 2012, doi: 10.1016/j.ijengsci.2011.08.010.

[6] Kanaun S and Tkachenko O, "Effective conductive properties of open-cell foams," *Int. J. Eng. Sci.*, vol. 46, no. 6, pp. 551–571, Jun. 2008, doi: 10.1016/j.ijengsci.2008.01.012.

[7] Akué Asséko A C, Cosson B, Duborper C, Lacrampe M F, and Krawczak P, "Numerical analysis of effective thermal conductivity of plastic foams," *J. Mater. Sci.*, vol. 51, no. 20, pp. 9217–9228, Oct. 2016, doi: 10.1007/s10853-016-0161-8.

[8] Doermann D and Sacadura J F , "Heat Transfer in Open Cell Foam Insulation," *J. Heat Transf.*, vol. 118, no. 1, p. 88, 1996, doi: 10.1115/1.2824072.

[9] Chiappini D, "Numerical simulation of natural convection in open-cells metal foams," *Int. J. Heat Mass Transf.*, vol. 117, pp. 527–537, Feb. 2018, doi: 10.1016/j.ijheatmasstransfer.2017.10.022.

[10] Li Z, Xia X, Li X, and Sun C, "Discrete vs. continuum-scale simulation of coupled radiation and convection inside rectangular channel filled with metal foam," *Int. J. Therm. Sci.*, vol. 132, pp. 219–233, Oct. 2018, doi: 10.1016/j.ijthermalsci.2018.06.010.

[11] Marvi-Mashhadi M, Lopes C S, and LLorca J, "Modelling of the mechanical behavior of polyurethane foams by means of micromechanical characterization and computational homogenization," *Int. J. Solids Struct.*, vol. 146, pp. 154–166, Aug. 2018, doi: 10.1016/j.ijsolstr.2018.03.026.

[12] Marvi-Mashhadi M, Lopes C S, and LLorca J, "Effect of anisotropy on the mechanical properties of polyurethane foams: An experimental and numerical study," *Mech. Mater.*, vol. 124, pp. 143–154, Sep. 2018, doi: 10.1016/j.mechmat.2018.06.006.

[13] Obi B, *Polymeric Foams Structure-Property-Performance: A Design Guide*, 1 edition. Waltham, MA: William Andrew, 2017.

[14] "EN 14315-1 Produits isolants thermiques destinés aux applications du bâtiment - Produits en mousse rigide de polyuréthanne (PUR) ou de polyisocyanurate (PIR) projetée, formés en place."

[15] Bargmann S *et al.*, "Generation of 3D representative volume elements for heterogeneous materials: A review," *Prog. Mater. Sci.*, vol. 96, pp. 322–384, Jul. 2018, doi: 10.1016/j.pmatsci.2018.02.003.

[16] Gibson L J and Ashby M F, *Cellular Solids: Structure and Properties*, 2 edition. Cambridge: Cambridge University Press, 1999.

[17] Doutres O, Atalla N, and Dong K, "Effect of the microstructure closed pore content on the acoustic behavior of polyurethane foams," *J. Appl. Phys.*, vol. 110, no. 6, p. 064901, Sep. 2011, doi: 10.1063/1.3631021.

[18] Sanchez-Palencia E, "Comportements local et macroscopique d'un type de milieux physiques heterogenes," *Int. J. Eng. Sci.*, vol. 12, no. 4, pp. 331–351, Apr. 1974, doi: 10.1016/0020-7225(74)90062-7.

[19] Nemat-Nasser S and Hori M, *Micromechanics: overall properties of heterogeneous materials*. Amsterdam: North Holland, 1993.

[20] Oudet C, *Polymères: structure et proprietés, introduction*. Masson, 1994.

[21] El Moumen A, Kanit T, Imad A, and El Minor H, "Computational thermal conductivity in porous materials using homogenization techniques: Numerical and statistical approaches," *Comput. Mater. Sci.*, vol. 97, pp. 148–158, Feb. 2015, doi: 10.1016/j.commatsci.2014.09.043.

[22] Lewis R W, Huang H C, Usmani A S, and Cross J T, "Finite element analysis of heat transfer and flow problems using adaptive remeshing including application to solidification problems," *Int. J. Numer. Methods Eng.*, vol. 32, no. 4, pp. 767–781, Sep. 1991, doi: 10.1002/nme.1620320408.

[23] Ahmad S, Irons B M., and Zienkiewicz O C, "Theory Reference for the Mechanical APDL and Mechanical Applications," p. 1226, 2009.

[24] Auriault J L and Caillerie D, "quelques remarques sur les méthodes d'homogénéisation sorne rernarks on homogenization processes," *Rev. Fr. GÉOTECHNIQUE*, p. 8.

[25] Auriault J L, *Milieux Hétérogènes Homogénéisation*. 2004.

AUTHOR INDEX